D0722490

THE REAL PLANET OF THE APES

THE REAL PLANET OF THE APES

A NEW STORY OF HUMAN ORIGINS

DAVID R. BEGUN

PRINCETON UNIVERSITY PRESS
PRINCETON AND OXFORD

Published by Princeton University Press, 41 William Street,
Princeton, New Jersey 08540
In the United Kingdom: Princeton University Press, 6 Oxford Street,
Woodstock, Oxfordshire OX20 1TW
press.princeton.edu
Jacket art: *Dryopithecus Africanus* (pencil on paper), English School,
(20th century) / Private Collection / Bridgeman Images

ISBN 978-0-691-14924-0
Library of Congress Control Number 2015947083
British Library Cataloging-in-Publication Data is available
This book has been composed in
Sabon Next Pro Lt & Grotesque MT Std
Printed on acid-free paper. ∞
Printed in the United States of America
1 3 5 7 9 10 8 6 4 2

CONTENTS

PREFACE

For the past thirty years or so I have lived in a world I like to call the real planet of the apes. It's not a New York or San Francisco overrun by bipedal talking chimps, gorillas, and orangs with weapons and on horseback, like in the movies, but instead, a large part of the world as it was millions of years ago. Believe it or not, between about 7 and 22 million years ago the planet was full of apes, or at least the Old World part of it (Europe, Asia, and Africa.) We know of about fifty species of fossil apes that roamed the forests of the Old World during that time span, which is called the Miocene. There were probably several times more than this number that may or may not ever be found. I find this amazing considering that there are only a handful of apes left today and they are disappearing fast. Aside from humans (we are in a real sense just bipedal apes), living apes are confined to patches of forest in Southeast Asia and central Africa. They are rare, all endangered in fact, but they have an amazing array of adaptations. Some are very big, others quite small, some dwell in trees, others mostly on the ground, some eat fruit, others leaves, and some are social, others solitary.

I want to know how we went from dozens of ape species at any one time, hundreds in all over the millennia, to just a few today. I want to know how apes evolved all of the unique adaptations they share. Why do I want to know? Because I know that without these adaptations humans would not exist.

Why are apes so rare today? How did they get to be the way they are? And, even more important to me, why do they resemble us in so many ways? We apes all have big brains, highly dexterous hands, mobile shoulders and hips, broad flat chests, and many other more subtle, shared similarities. As a group we are the most intelligent animals on the planet (except possibly dolphins). There is a reason for this—and that reason is bound up in the fossil record of the apes, waiting to be understood. I attempt to answer these questions in this book, based on my interpretation of the evidence, which I have been trying to make sense of these last three decades.

I try to share my passion for paleoanthropology in this book, along with what I hope is a logical narrative, starting with ape origins and ending with the origins of humans, by which I mean the human lineage after chimps and we diverged. Along the way I want to share some of my experiences with all the amazing people I have met, as well as my discoveries and screwups. I have tried to point out areas of disagreement or controversy where my ideas do not exactly follow those of other colleagues.

Few of the ideas and interpretations expressed in this book are exclusively my own but, rather, are the result of more than a century and a half of research. The few that are mine more or less exclusively tend to be controversial. One in particular is my conclusion that it is Europe, and not Africa, that is the center of origin of the ancestors of living great apes and humans. The conventional wisdom is that the common ancestor of all living great apes and humans evolved in Africa. After all, chimps and gorillas, the closest relatives of humans, live in Africa today, and the earliest human ancestors are also African. The idea of an African origin dates back to Darwin, but even the great man himself had his doubts, as we shall see. Some researchers like my idea and some do not, for reasons that will become clear later. I obviously like the idea myself, but it is just a hypothesis, not a conviction. Anyone who thinks they know exactly and without a doubt what happened 10 million years ago, or for that matter at any time during the evolution of life on Earth, needs to reconsider their views. I try to keep an open mind, but I have come to a certain number of conclusions that are currently consistent with the evidence, at least as I see it. All that might change, and that is exactly the challenge—coming up with ideas that can be falsified with new evidence. I hope my friends and colleagues and readers in general will see that this has been my goal.

• • •

When I told my parents that I was quitting the premed program at Brandeis University to go to the University of Pennsylvania and study human evolution, my father said "Human evolution. A good hobby for a doctor." Fortunately my folks warmed up to my interests; in fact, they were always completely supportive, and I thank them for that. It may have been inevitable that I would turn to paleoanthropology, having spent all my summers from birth to nearly age 20 with my grandparents in the French region of the Périgord, the birthplace of prehistory. The earliest modern humans from Europe are often referred to as Cro-Magnons. The original Cro-Magnon fossils were found about forty kilometers from my grandparents' home. The word "Cro-Magnon" is Occitan (or Patois), the local language my grandfather spoke before he learned French. Edouard Lartet and his son Louis Lartet, working in the Périgord, essentially discovered prehistory. Earlier and further south, Lartet the elder had described the first fossil primates, *Pliopithecus* and *Dryopithecus*, two creatures that have kept me busy for thirty-plus years. The world famous painted cave site of Lascaux, which I visited many times as a child, is also in the Périgord.

I am immersed in the culture of the Périgord, which is as much about prehistory as it is about gastronomy (foie gras and truffles) and castles. My grandfather, who received the French Legion of Honor for service in the French Resistance during the Second World War, was a personage in the Périgord. He knew the prehistorians, whose sites he helped electrify as director of the EDF (Electricité de France) in the Périgord. My grandparents and my mother became close with Josephine Baker, the jazz performer, civil rights activist, and member of the French Resistance during World War II, whose castle in the Périgord my grandfather helped to restore. Along with my grandparents, my French mother and my American father, a soldier who married a French girl, instilled in me a love for the Périgord and inadvertently for prehistory. I have no doubt that my emotional attachment to this area of the world influenced the course of my life and my choice to become a paleoanthropologist.

I left a premed program to go to Philadelphia and work with Alan Mann, given his interest in middle and late Pleistocene humans, such as Neandertals and their ancestors. I was sure, as sure I had been earlier that I wanted to be a doctor, that I wanted to study Neandertals. Not long before I came to Penn, Alan had made molds of Neandertal and *Homo heidelbergensis* fossils in Central Europe. While in Hungary he also, almost incidentally, molded specimens from Rudabánya, a late Miocene fossil great ape site from which a new genus of fossil ape, *Rudapithecus*, had been identified. Among the fossils that Alan had molded were specimens, for the most part phalanges (finger bones), that were undescribed. I jumped at the opportunity to work on original undescribed fossils and started my life with the "real planet of the apes": the Earth during the Miocene. A fortuitous if not spastic jump from an interest in 200,000-year-old Neandertals to 10-million-year-old apes. I am still amazed and grateful to all the senior scientists who received me in their labs and even in their homes, starting when I was barely twenty-one and really had no idea what I was doing.

The point of this reminiscing is to document the degree to which life takes unexpected twists and turns. The one constant is a passion for the discipline. When I think back on it, there is so much that is random and unpredictable. I tell students to just go for it, be serious and dedicated and try to be as confident and humble as you can manage. To all of those parents out there who agree with my father that paleoanthropology is a good hobby for a doctor, I apologize if I have in any way influenced your children toward this direction in life.

THE REAL PLANET OF THE APES

INTRODUCTION

For more than 150 years now it has been widely accepted among biologists and most other scientists that humans evolved from an ape. Not one that lives today, such as the chimpanzee, our closest living relative, but one that lived millions of years ago. Let me make that perfectly clear. We did not evolve from a chimpanzee, nor did chimps evolve from us. Rather, chimps and humans evolved from an unknown ape that lived before humans and chimpanzees branched off from each other, at least 7 million years ago, to pursue their own evolutionary destinies. What was this ape like? And from what did it evolve? And further back in time, what was the common ancestor of all apes and humans like? And what might studying ancient apes tell us about what makes us human?

These are the questions that have been driving my research since I was a graduate student in the 1980s and that motivate this book. We have known, ever since the work of Charles Darwin and other researchers of the late nineteenth century, that we have a special evolutionary relationship with apes. In a real sense, we *are* bipedal apes. Darwin and his principal defender, Thomas Henry Huxley, concluded in the 1870s that humans not only share a common ancestor with apes, we share a common ancestor with African apes to the exclusion of all other primates. To put this another way, Darwin and Huxley concluded that chimpanzees and gorillas are more closely related to humans than to orangutans. Orangutans are Asian great apes, a branch of the great apes that split from the common ancestor of the African apes and humans.

Among Darwin's contemporaries and researchers well into the twentieth century, the idea that African apes could be more closely related to humans than they are to orangs was hard for most scientists to swallow. Many rejected this conclusion, favoring the hypothesis that humans branched off first, and then what we call the great apes—chimps, gorillas, bonobos, and orangs—went off in their own different directions. After all, humans are very different from the apes, physically and mentally. Great apes all look more or less the same, at least superficially. They are large, hairy beasts with long arms, short legs, big jaws and teeth, and small brains compared with humans. Still, all of these early researchers recognized the reality of human evolution and our connection with apes. As we will see, there are in fact many differences among the great apes, and as a group they are not as different from humans as they may first appear.

Today, scientists group modern humans with the great apes. Although we have bigger brains and walk upright on our hind limbs, we share an enormous amount with the apes, from an almost indistinguishable genome to more similarities in our structure and behavior than we share with any other living organism. This is the reason that scientists group us with apes, to the exclusion of all other organisms, in the superfamily Hominoidea. We are especially closely related to the great apes, and so we share with them a place within the family Hominidae. (Figure 0.1 illustrates the relationships among the apes discussed in this book.) We will learn more about how all primates, apes, and humans are related later in this chapter.

In this book, I want to tell the story of ape evolution over approximately 30 million years. In many ways, my account differs from the usual textbook account, but I think it better explains how we got to be the way we are. Indeed, it is impossible to understand and explain the course of human evolution without understanding ape evolution first. Modern human anatomy can only be understood as a direct consequence of having evolved from an ape. We did not evolve from any living ape, but the anatomy of the common ancestor we share with apes sets the stage for human evolution. Everything from the structure of our teeth to our brains, our dexterous hands, our upright posture, and even our reproductive biology all have precursors in the anatomy of our ape ancestor. Many of these attributes are still found in living great apes, which is one of the reasons we study them so thoroughly.

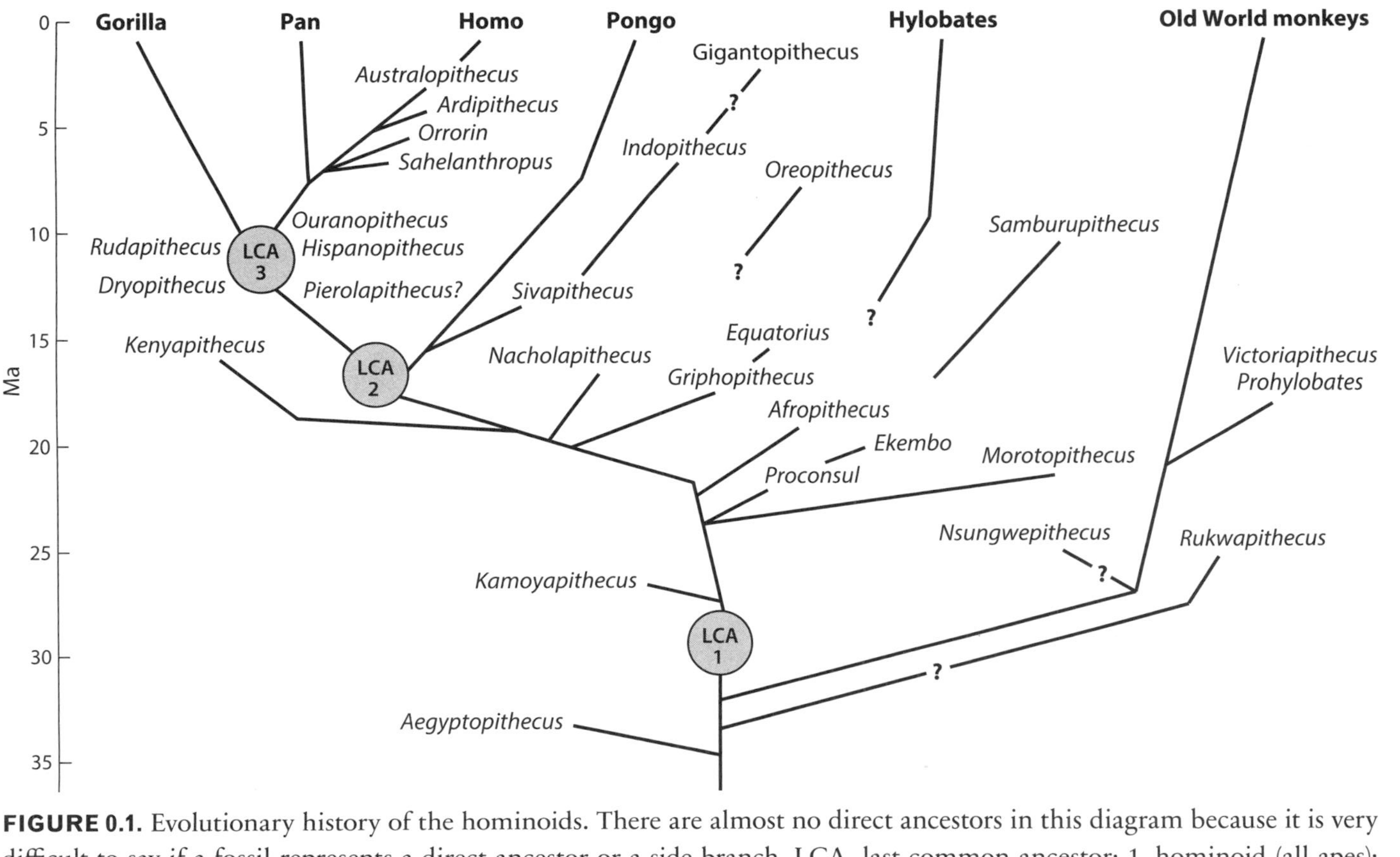

FIGURE 0.1. Evolutionary history of the hominoids. There are almost no direct ancestors in this diagram because it is very difficult to say if a fossil represents a direct ancestor or a side branch. LCA, last common ancestor; 1, hominoid (all apes); 2, hominid (great apes and humans); 3, hominine (African apes and humans). The fossil ape *Oreopithecus* and *Hylobates* (gibbons and siamangs) cannot be linked confidently to any known branch of the hominoid evolutionary bush. *Nsungwepithecus* may be the oldest known cercopithecoid (Old World monkey).

When I began my study of fossil apes from Europe, nearly all of them were attributed to the genus *Dryopithecus*, an animal scientists considered to be far removed from the central court of great ape and human evolution—a side branch in early ape evolution. I found no reason to challenge this conclusion at the time. However, I did come to the conclusion as I was completing my thesis that these fossil apes are great apes. In other words, they are hominids (the group that includes orangs, gorillas, chimps, bonobos, and humans.) With the discovery of more fossils, especially from Spain but also from Hungary, the view that European apes are more central to the question of hominid origins became more widely accepted. Before this, most scientists believed that European apes were just an interesting side story. Fossil apes were better known at the time in Africa, and as I said earlier, the overwhelming consensus was that great ape and human evolution was a mainly African story.

As I continued to work on European apes I had another idea. These fossil apes are not only great apes but also African apes (hominines.) This was a new idea—and it is not as widely accepted today as the conclusion that European apes are great apes. The reason this new idea is controversial is the same old story—every event of any significance in the evolutionary history of apes and humans was widely considered to have occurred in Africa. It took many new discoveries and detailed research on these new specimens to build the case for the presence of African apes in Europe.

Nevertheless, I began to wonder what African apes were doing in Europe 9 to 12.5 million years ago. At first I thought that they had dispersed into Europe from Africa, a sort of excursion by a side branch of African apes that eventually led to extinction. Finally it occurred to me that the ancestor of the African apes and humans may actually have evolved in Europe instead of Africa. While it was a radical departure from widely accepted reconstructions of ape and human evolutionary history, I was intrigued by the African ape features I found in European fossil apes and the complete lack of evidence for fossil great apes in Africa during the same time period. I found evidence that European Miocene apes were more advanced (or derived, in scientific terminology) than apes from the early Miocene of Africa and share characters with living African apes and humans. I hypothesized that African apes evolved in Europe and moved to Africa, not the other way around. Not to overdramatize the point, but I do think of this as a eureka moment that has to some extent defined the trajectory of my career afterward.

I am determined to falsify this hypothesis. That may sound strange. But we cannot really prove anything in paleontology. We can only try to find evidence inconsistent with prevailing theories. The only way to formally test my hypothesis is to seek to disprove it. To do so, I have been testing this hypothesis with new specimens. Many European fossil apes share characteristics of orangutans, which I interpret as primitive (they evolved first). But they all have features of African apes, as we shall see later. As surprising as it is, there is strong evidence, which I will reveal, to support my hypothesis. This is, to be honest, a hypothesis that goes against most opinion all the way back to Darwin, though as we shall see, Darwin was more open-minded than many persons are today (He was receptive to the idea that *Dryopithecus* might have a connection to African apes). To summarize the major conclusions of this book, my research and that of many colleagues has led me over the last thirty years to a number of conclusions. Apes evolved in Africa from ancestors of African origin (e.g., *Aegyptopithecus*). By about 20 million years ago, primitive apes, more monkey-like than apelike, began to flourish in Africa. Among these apes, one was better equipped to disperse to more seasonal climates (Eurasia). This ape, with its abilities to exploit a broader range of resources than the first apes, was poised to take on Eurasia. The ecological conditions in Eurasia selected for new adaptations in apes. The ape that dispersed into Eurasia began to evolve novel features related to diet and positional behavior, eventually splitting into the two major groups of living great apes, the pongines in Asia and the hominines in Europe. At the same time, the fossil trail in Africa went cold temporarily (there are no fossils), while in Eurasia ecological conditions favored further changes in locomotion (suspension) and increases in brain size. Large-brained and suspensory apes flourish until a progressively cooling and drying Eurasia eventually caught up with them. Many went extinct but a few were able to disperse south, tracking the forests retreating toward the equator. The ancestors of orangutans ended up in Southeast Asia while the common ancestor of the African apes and humans settled somewhere in the African tropics. By roughly 10 million years ago, gorillas separated from the common ancestor of chimps and humans, and by about 7 million years ago, chimps and humans diverged. It is at that point that the human lineage emerged. If you find the narrative confusing as you read the book, don't give up. Come back here and remind yourself of the major events in ape and early human evolution. It is a story, and I hope it will make sense to all of you. Before we embark on this grand narrative, it is important to learn more about primates and apes, and crucially what characteristics we humans share with them.

HUMANS ARE PRIMATES

We belong to the zoological order known as the Primates. The classification of primates can be complicated, so I will make it simple here. Primates are traditionally divided into prosimians and anthropoids. Prosimians include lemurs, lorises, and galagos or bush babies, among others, and tarsiers.[1] Anthropoids include New World monkeys, Old World monkeys, apes, and humans. Nearly all anthropoids are daytime active and most are larger than prosimians. They have larger brains and rely less on insects and more on fruit and leaves. Most anthropoids are highly social and even more visually oriented than prosimians. Most anthropoids are also tree dwelling, but some spend time on the ground, especially the larger Old World monkeys (baboons) and African apes.

All primates are intelligent, dexterous, clever, vision-oriented animals, mostly tree dwelling, with grasping hands and feet (except humans; our feet have lost this ability). Brain size and eye-hand coordination is generally greater than in other mammals of similar size. It is clear that the evolution of the primates forms the foundation for the evolution of the apes and humans.

Most people use the word "monkey" to refer to those hairy, four-legged critters that kind of look like us and can be trained to do clever things. It is common to refer to chimps or gorillas as monkeys, but in fact, apes and monkeys are very different. First of all, there are two groups of animals that biologists call monkeys. One of them, the New World monkeys, live, as the name suggests, in the New World (Mexico and Central and South America but not the United States or Canada). The other kind of monkey lives in Africa and Asia, and we call them the Old World monkeys. While we refer to both groups colloquially as monkeys, New World monkeys are distinct and evolved separately from the Old World monkeys, apes, and humans. So, from now on, when I refer to monkeys, I am talking about Old World monkeys (see plate 1).

Old World monkeys, apes, and humans all fall within the zoological category of the catarrhines. New World monkeys are in a different group, having diverged before Old World monkeys and apes split. Roughly 35 million years ago, there was a population of primates that gave rise to the catarrhines to the exclusion of all other primates. In other words, we and the other catarrhines branched off from the common ancestor of the New World monkeys at that time and have since evolved in our own way. Since that initial branching event, Old World monkeys and apes have branched off from one another, as have each of the lineages of living apes and humans in turn.

Old World monkeys include baboons and macaques, which are the most common, but many other species exist. Monkeys are intelligent and very flexible in their behavior. These traits allow them to adapt to life in harsh climates, such as the snowy mountains of Japan, though most species, like the apes, live in the topics or subtropics. They generally live in social groups with complex hierarchies in which rank is important and can be inherited. The offspring of a high-ranking monkey are likely to be high ranking, too. Monkeys almost always give birth to one infant per pregnancy and invest a great deal of time raising and nurturing their young.

Compared with other mammals of similar size, monkeys generally have larger brains. They are very agile, and while some are more adapted to an arboreal (tree-living) lifestyle and others are more terrestrial (ground-dwelling), all monkeys move quickly and adeptly both on the ground and in the trees. Stay clear of them if you go on a safari or visit a place where they run free; their antics may be cute, but many have lost their fear of humans and they will not hesitate to bite. In fact, in the West we tend to think of monkeys as adorable fuzzy creatures, but where they live side by side with humans, they are often not very well liked, mainly because they steal food, destroy property, and raid crops.

So monkeys are intelligent and adaptable, traits that have allowed them to thrive alongside humans. It is no exaggeration to say that monkeys are among the most intelligent creatures on the planet. However, compared with monkeys, apes are in another category altogether.

Most of the apes are much larger in both body size and brain size than monkeys and all apes lack tails. (I am talking about living or extant apes here; some of these attributes were not present in the earliest apes, as we shall see.) Their arms are long (longer than their legs), allowing them to swing below the branches, whereas monkeys mostly walk along the tops of the branches. Female apes go through a menstrual cycle rather than a more seasonal reproductive cycle (estrous), as female monkeys do. Apes live longer and have larger, more complex societies and more complex and social brains. Apes score higher than any monkey in lab tests of intelligence, and some researchers have even claimed to have been able to teach apes to communicate in a rudimentary language. Although there are detractors regarding the ape language experiments, it is clear that apes are capable of much more cognitively challenging tasks than monkeys. In the wild, great apes, especially chimpanzees, make and use tools in foraging. All great apes build nests to be comfortable and safe at night in the trees or on the ground, they make umbrellas and other devices to protect themselves from the elements, and they devise novel and intelligent solutions to the problems they face. In my opinion, ape intelligence, specifically great ape intelligence, is an order of magnitude above that of any monkey and makes an obvious comparison with early humans. That is really the main reason we study great ape behavior in the wild. They surely represent something close to the way the earliest humans, which were great apes themselves, behaved. We did not evolve from a living great ape, but the earliest human species anatomically resembled living great apes and surely behaved much more like living great apes than living humans. Why we have changed so much and apes so little is possibly the biggest puzzle in paleoanthropology.

Scientists divide the living species of apes into two groups. Lesser apes include the gibbons and a less familiar primate called the siamang. Gibbons and siamangs are placed in the family Hylobatidae, or hylobatids, more informally. As we learned earlier, the great apes include chimpanzees, bonobos (formerly known as pygmy chimps), gorillas, orangutans, and humans. They are collectively known as the Hominidae, or hominids. The vast majority of paleoanthropologists recognize this dichotomy and are okay placing humans and great apes in the same family (hominids), though there are some who continue to elevate humans to a unique family (hominid) and place great apes in another group called the pongids. This is an evolutionarily artificial classification that places the unique adaptations of humans above the genetic and anatomical evidence of our relationships with the great apes. It runs counter to normal taxonomic practice and, I would say, plays into the hands of those who refuse to accept that humans evolved from an apelike species, or evolved at all. Here I use "hominids" to refer to the great apes and humans (see plate 2).

The recognition of the similarities between great apes and humans is remarkable and recent. Up until the 1990s most researchers reserved the word "hominid" for humans and our ancestors—all species on our branch of the family tree after we split off from the last common ancestor we share with a great ape. In most of my publications through the mid-1990s, I had to justify the use of the term "hominid" as employed here as if it was controversial and confusing. Today, reserving the term "hominid" for humans has become controversial and confusing. Most researchers, myself included, place African apes (chimpanzees, bonobos, and gorillas) and humans (we and our fossil relatives) in the same subfamily, the Homininae (hominines) (table 0.1). A few researchers even advocate including the genus of chimpanzees and bonobos (*Pan*) within our genus *Homo*, as a subgenus. Most researchers today agree that chimps and gorillas are more closely related to humans than they are to orangutans. Furthermore, they agree that chimps are more closely related to humans than chimps are to gorillas. In a family

TABLE 0.1. A Classification of Living Hominoids.

SUPERFAMILY	FAMILY	SUBFAMILY	TRIBE	GENUS
Hominoidea				
	Hylobatidae (gibbons and siamangs)			
				Hylotbates
				Nomascus
				Hoolock
				Symphalangus
	Hominidae (great apes and humans)			
		Homininae (African apes and humans)		
			Hominini[1]	
				Homo
				Ardipithecus
				Australopithecus
			Panini	
				Pan
			Gorillini	
				Gorilla
		Ponginae (orangutans)		
			Pongini	
				Pongo

[1]The hominini includes additional taxa not included here, for clarity.

tree, chimps are our sisters, gorillas our cousins, and orangs are our cousins once removed.

WHAT MAKES AN APE?

Earlier I listed a few broad characteristics of biology and behavior that distinguish monkeys from apes and humans. The distinctions are important because in a way they approximate the path of evolution from our more monkey-like ancestors to our apelike ancestors to us. It is important to remember again that none of the living primates are our ancestors and that they have all evolved their own special characteristics. But we can see in monkeys today the enhanced intelligence, adaptability, and agility that was present in our common ancestor with them. We can see in the apes the further development of the brain and changes in body plan. Looking at the differences between moneys and hominoids helps us to retrace, as an approximation, the course of our evolution. So let's look at this distinction in more detail.

First, there is the genetic evidence. We have known since the early part of the twentieth century that humans and apes are more similar to each other than to any other primate, when various organic molecules are compared. At first, comparisons among monkeys, apes, and humans involved proteins in the blood. Over a century ago researchers began to document similarities among Old World anthropoids to the exclusion of other primates.

In 1901 George Nuttall published a study in which he described a blood test to assess relationships among animals. Nuttall injected human blood serum (blood plasma from which the fibrogen, a clotting protein, has been removed) into rabbits. Blood collected from the rabbits was used to create an antiserum for human blood, that is, rabbit plasma with antibodies to human blood serum, produced by the rabbits as a natural immune reaction to human blood. Now he had a substance that he could use to detect human or humanlike blood. When Nuttall mixed this antiserum specific to humans with the sera from hundreds of different animals, he found that almost none of the mixtures reacted. In other words, there was nothing in the sera of any of these other animals that the human serum antibodies would react with—except monkey sera. The antibodies to human serum recognized something in the monkey sera.

Nuttall followed up this work in 1904 with a monograph entitled "Blood immunity and blood relationship: a demonstration of certain blood-relationships amongst animals by means of the precipitin test for blood." He concluded that humans share a close relationship with the great apes and that the next closest relatives, in order, were Old World monkeys, New World monkeys, and prosimians (lemurs, lorises, galagos, and tarsiers). This was really a remarkable conclusion for the time because it is essentially what we think today. Nuttall even suggested that, given the difficulty of finding informative fossils, molecular techniques might be the best way to classify species and determine their evolutionary relationships.

Whether relationships among species can be determined by morphology (which is essentially all we have for fossils) or whether genes are the only reliable source of information is a debate that rages on today. As a paleoanthropologist, I advocate using morphology, as well as molecules, to help unravel the mysteries of ape and human origins, and I think that most of my colleagues would agree—even the molecular systematists who use DNA to reconstruct the tree of life.

As techniques grew more refined, it became possible to begin to differentiate among the Old World anthropoids. In the 1960s, researchers first proposed, based on molecular evidence, that humans are specifically related to African apes (this of course had been concluded much earlier by Darwin, and especially by Huxley, from morphological evidence).

Today it is possible to make detailed comparisons among organisms based on the actual sequence of base pairs in their DNA. The vast majority of analyses comparing the DNA sequence of humans with those of other primates yield the same results. Humans share a most recent ancestor with chimpanzees and bonobos. Gorillas are next most closely related to humans and chimps, and orangs are the next group out. Gibbons and siamangs are the so-called lesser apes, the living hominoids that first branched from the line leading to the great apes and humans.

Let's put this in context.

Figure 0.2 is a diagram called a cladogram, which shows a nested set of relationships. It depicts the order of branching events but not direct ancestor-descendent relationships. You should read the cladogram as follows: New World monkeys branch off from the line leading to Old World anthropoids (catarrhines) and each goes on their own evolutionary path. This means that all New World monkeys have the same evolutionary relationship to all catarrhines. While this may seem counterintuitive (after all, we speak of both Old and New

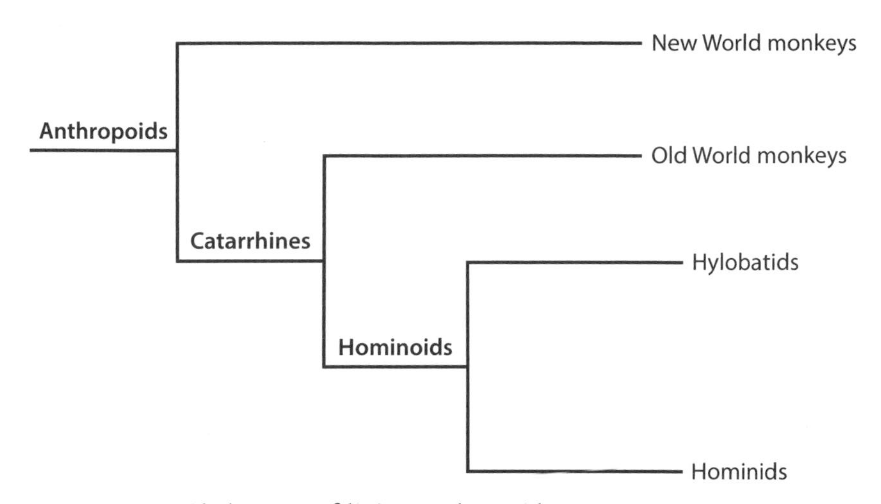

FIGURE 0.2. Cladogram of living anthropoids.

World monkeys), it is in fact documented by many lines of evidence, both morphological and molecular. So, New World monkeys as a group all have the same relationship with catarrhines (Old World monkeys, ape, and humans.) Both groups branched off from each other. Although capuchin monkeys from South America look a lot like vervet monkeys from Africa, they are in fact no more closely related to vervets than they are to Elvis (or any other hominoid). Vervets are more closely related to Elvis and all other hominoids, as well as to all other Old World monkeys, than they are to capuchins and all other New World monkeys. Get it? If not, have another look. It takes a while to properly understand a cladogram.

The next branching event separates the Old World monkeys from the apes (including humans). Again, this means that all Old World monkeys have the same evolutionary relationship to all hominoids. Although they may look to the lay person more like Old World monkeys than like humans, gibbons (hylobatids) are in fact more closely related to humans than they are to Old World monkeys.

And so it goes down the line. Gibbons branch off from the common ancestor of great apes and humans, and within this group, orangs diverge from African apes and humans, and finally gorillas diverge from chimps and humans. Despite the behavioral and morphological similarities that exist today among the African apes, chimpanzees and bonobos are more closely related to humans than they are to gorillas. This fact is reflected in the extreme similarity of the DNA of chimpanzees and humans, in the chimp-like anatomy of our ancestors, and in details of anatomy shared by humans, especially fossil humans, and chimps today (from now on, when I say "chimps" or "chimpanzees," I am including bonobos as well.) The reason that chimps and gorillas look more similar to one another than chimps resemble humans, even though chimps and humans are more closely related to one another, is that chimps and gorillas share primitive characters that humans have lost. Our earliest fossil ancestors looked much more like chimpanzees than we do today.

Let me return to the cladogram and the science behind it, cladistics. Why is it that evolutionary biologists do not simply rely on overall similarity in reaching their conclusions about evolutionary relationships? After all, geneticists determine relationships based on overall similarity of the genomes among species. The problem with morphology is that it is generally limited to a relatively small number of characters compared with the huge number of genes included in a genetic analysis. And anatomical characters do not all have the same usefulness for working out evolution in given lineages.

Species share anatomical features for three main reasons. They may have evolved in an ancient common ancestor shared by many other species. We have four limbs, like turtles, lizards, crocodyles, birds, and mammals, among other animals. But this does not mean that lizards and humans are more closely related to each other than humans are to dolphins, which lack externally visible limbs. It just means that lizards and humans evolved from a common ancestor that had four limbs (tetrapods). During the time since that ancestor lived, some descendants have lost their limbs (dolphins and whales; snakes). These shared characters (for example, having four limbs) are called primitive characters, and they are not helpful in deducing relationships among closely related species. Small brains (compared with humans'), long arms, short legs, large canines, and a hirsute appearance are all primitive characters that make chimps and gorillas resemble one another but do not tell us that they are most closely related. For much of our evolutionary history we looked like that as well.

We know that lizards and snakes are more closely related to each other than either one is to turtles or humans because they share features inherited from their last common ancestor, an animal that lived after turtles branched off but before mammals appeared on the scene. In the same way, humans and dolphins share many features with each other that nonmammals do not have because they were inherited from the common ancestor of dolphins and humans, an early mammal. Characters inherited from the last common ancestor of a group of organisms are referred to as derived characters. These are the characters that we need for establishing evolutionary relations.

I mentioned that there are three main ways in which characters can be shared among species. The third process that can lead to shared characters is parallel evolution. I will discuss this in more detail later, but for now let's just say that it is the independent evolution of similarities. In primate evolution, adaptations allowing species to hang below branches evolved independently quite a few times: in lemurs, New World monkeys, and maybe even more than once in apes. Thickly enameled teeth set in massive jaws also evolved multiple times in apes. We will explore the reasons for these fascinating parallel events later in the book.

The difficulty that some people have in accepting the idea that chimpanzees could be more closely related to humans than to gorillas, or that we could be related in any way to apes, is really an artifact of our perceptions of humans as apart from the other animals. We all know that humans are animals (as opposed to plants or fungi or bacteria), but we often separate the two categories in our minds. Many religions also teach that humans are separate from animals and have a special and unique origin. So our understanding of a chimpanzee is biased by the influence of our cultural traditions. When we consider them outside of this frame of reference (as much as we can), it becomes obvious that chimps and humans share a special, extremely close relationship.

We've known since the work of Mary-Claire King and her dissertation supervisor, Allan Wilson, at Berkeley in the early 1970s that chimpanzees and humans share almost 99% of their DNA in common. However, this also needs to be placed in context. It is estimated that there are between about 20,000 and 30,000 genes in the genome of every mammal. In the case of humans and chimpanzees, when comparing the same gene, there is on average about a 98.8% similarity in the base pairs. Base pairs are the nucleotide pairs that make up the rungs of double helix molecule. But humans and chimps have about 3.3 billion base pairs, so when you multiply the relatively small difference between the genomes of chimps and humans by 3.3 billion, the estimated number of genetic differences between humans and chimps adds up to quite a few, about 40 million, in fact. While many of these base pairs do not contribute to the functional portion of the genome (genes), the number of differences among genes is still large. Scientists also estimate that a good number of genes are unique to chimpanzees and others are unique to humans, adding to the difference. Furthermore, it is likely that many of the genes that distinguish humans and chimps are regulatory genes, which have multiple so-called downstream effects. A few regulatory gene differences can probably make the difference between a chimp and a human pattern of growth and development. So, although 98.8% sounds like a small difference, given the number of genes in the chimp and human genomes and the effect of regulatory genes, it amounts to more than enough genetic divergence to account for the differences in biology and behavior between chimps and humans.

While chimps and humans are genetically closer to one another that either is to any other primate, all hominoids share the vast majority of their DNA sequence. In fact, all vertebrates share most of their DNA. It is estimated that we have 88% of our genes in common with mice, 65% with chickens, and even 25% with grapes. For the most part, the same genes make all organisms work the way they are supposed to, and if it ain't broke, don't fix it. However, all hominoids share a range of anatomical characteristics that distinguish them from other primates. The list is long, so I will focus on the major features.

A larger brain. Once body size is taken into account, all hominoids have larger brains, on average, than other primates. If you were to make a comparison between the brain and body mass ratios of humans and mice, you would find that the mouse has a brain about the same size as humans relative to overall body mass, roughly 2%. So it sounds as if mice have brains comparable with those of humans.

The problem with this sort of comparison is that it does not take into account the fact that different attributes of the body change size and shape at different rates, a phenomenon known as scaling. When you compare brain and body size ratios across a wide spectrum of mammals (the so-called mouse-to-elephant curve), you see that overall body mass increases much more rapidly than brain mass (figure 0.3).

Elephants, for example, have brains that weigh less than 0.2% of their overall body mass. That is roughly 10 times smaller than for the mouse. The reason is that an elephant is not a gigantic mouse. The ancestor of all elephants was an animal the size of a rabbit that lived 60 million years ago. As the elephant grew in size during the course of its evolutionary history, its brain grew as well, but much more slowly. This is expected because brains are, metabolically speaking, extremely expensive, more so than any other organ. An elephant could not feed a brain 10 times the size of its actual brain, not to mention the problem of fitting such a large brain into the skull or getting a big-brained baby elephant through a birth canal.

So brains and body masses do not scale at a ratio of one to one. If you compare the average brain-to-body mass of a large number of mammals you see a trend, and you can use this trend line, or regression, to predict how big a "typical" mammal's brain should be for a given body mass. It turns out that primates in general have significantly larger brains than most mammals at the same body masses. Within primates, monkeys scale with larger brains than lemurs and lorises, and apes have the largest brains, with a few exceptions (for instance, gibbon brains are about the same size as baboon brains after body mass is taken into account).

Among the hominoids, modern humans have much larger brains than a typical mammal of our body mass. For example, a fallow deer weighing about 70 kilograms (154 pounds) has a brain size of about 160 grams (5.6 ounces), while a human of the same body mass can easily have a brain 10 times that size. In great apes, the figure would be about 3.5 times the size of the deer's brain.

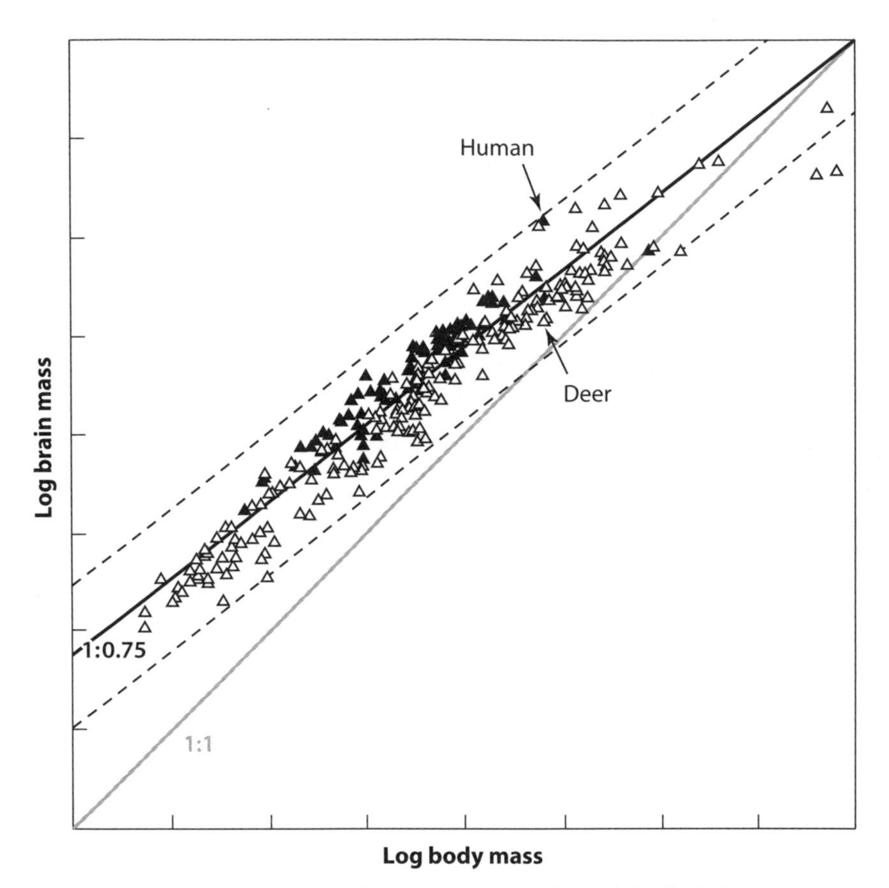

FIGURE 0.3. The mouse-to-elephant curve. The solid black line approximates the curve based on data gathered on the brain and body masses of hundreds of individual mammals. A slope of 1 (*gray line*), indicates that both variables are changing at the same rate. The actual line (solid black) is below 1 (0.75), indicating that brain size is increasing more slowly than body mass. The position of humans and the fallow deer show how much brainier we are; that is, how much larger our brains are than expected for animals of our size. The data point close to the human point (*black arrow*) is *Homo erectus*. (Modified from Martin 1990.)

We know that brain size is somehow related to information processing, or intelligence in some sense, but the relationship is not clear. In general, mammal species with larger brains tend to outperform those with smaller brains in various tests, but this link between brain size and cognitive performance is far from universal. *Within* species, however, there is no clear documented relationship between brain size and intelligence. In humans, for example, individuals of average or even exceptional intelligence and achievement in life can range in brain volume from under 1000 cubic centimeters to about 2000 cubic centimeters. The fascinating but somewhat morbid practice in the past of weighing, examining, and preserving the brains of renowned historical geniuses has shown that they can have brains anywhere in this range. Albert Einstein's brain is famously average in size (1230 grams upon his death in 1955).

In the lab setting, great apes routinely outperform monkeys in tests that require memory, recognizing objects, understanding symbols (including language), and recognizing their own reflection in a mirror. Great apes in captivity assemble objects to make compound tools. In the wild, great apes also make tools, with chimpanzees by far the most prolific and skillful tool makers. The social interactions among great apes are considered to be more complex than those of other primates, (the relationship between sociality and brain size isn't universally accepted) and can include recognition of roles and social status and anticipation of actions by other apes. Apes can also adjust their behavior based on a given situation. They may behave differently in a one-on-one situation than they would if a third ape were present.

One way apes use their larger brains is by developing what some researchers call culture, a concept that many anthropologists usually reserve for humans. However, wild chimpanzees have documented traditions that are passed along socially from one generation to the next and are unique to individual groups. Chimps in different populations have their own ways of making tools and their own sets of tools for specific tasks. The way a chimp uses stones to crack nuts in one area differs from the way stones are used in another. Hunting is relatively common in some chimp populations and rare in others. In other words, these chimps are carrying on traditions of learned behavior unique to their specific groups, just like humans. Bonobos, on the other hand, have not been observed to make tools in the wild, and when they do hunt, it is for small vertebrates such as frogs. However, in captivity bonobos are very proficient at making tools.

There are many other aspects of great ape behavior that are much more similar to human behavior than may appear at first glance. This is especially true for chimpanzees. In addition to making and using tools more frequently than other great apes, chimpanzees form coalitions, especially among males, to attack other groups or to defend themselves. They cooperate in hunting forays and, unlike other great apes, manage to capture and kill a variety of mammals, including red colobus monkeys and bush babies. While this behavior occurs occasionally in other primates, particularly baboons, it appears to be more common in chimpanzees. On the darker side, like humans, these coalitions of male chimps sometimes wage war on other groups; they deliberately and strategically kill, that is, murder, members of rival "gangs" and commit infanticide.

Female chimps also form coalitions; in fact, at many field sites where they have been studied, core groups of females appear to form the most stable components of chimp societies. Females do seem to compete among one another and are at least as antagonistic among themselves as are male chimps. Despite this competition and squabbling, females appear to be the glue that holds chimp societies together. I am not a psychologist, but my understanding of the research on great apes suggests that the way men and women behave socially is so similar to chimp behavior as to suggest that it is at least in part due to genes inherited from the common ancestor we share with chimps. The popular observation that men are from Mars and women are from Venus may in part be a legacy of the common ancestor we share with chimpanzees.

A straight spine and long arms. Below the neck we share many similarities with the other apes. All hominoids (apes including humans) have a relatively vertical posture compared with the Old World monkeys. Humans have a completely vertical posture, but gibbons and orangutans have vertical backbones as well, because they spend a great deal of their time hanging from branches. Apes have short lower backs compared with the rest of the vertebral column. This probably increases their stability in vertical positions. Their arms are so long that both gibbons and orangutans often walk upright on the ground or on the tops of branches with their arms held above their heads. Gorillas and chimpanzees have angled or obliquely oriented vertebral columns. Because their arms are longer than their legs, when they walk on all fours, their shoulders are above their hips. In fact, all nonhuman apes have arms that are much longer than their legs, unlike any other living primate (plate 3).

All apes except humans are exceptional climbers and tend to move along the branches from underneath rather than above. For the most part, monkeys move on the tops of branches. Apes swing arm-to-arm below the branches, which primatologists call suspensory locomotion, or brachiation. African apes spend much more time on the ground than the orangutan of Asia, but they are still very adept at swinging through the trees.

Shoulders for motion, hands for grasping. In addition to the structure of their vertebral columns and long arms, apes have a number of other characteristics that help them swing beneath the branches. Apes have broad chests with their shoulder blades positioned on the back, which moves their arms out to the side of their chests, rather than beneath their chests, as in monkeys and most other quadrupeds. Apes have highly mobile shoulder joints, which allow them to place their arms in many different positions, an ability that is critical to maintaining their balance in the trees. A fall could be fatal. While many other mammals spend most of their time in the trees, few are as large as apes.

All apes can straighten their arms completely at the elbow, and they all have flexible elbow and wrist joints that bend and rotate, giving them a wide range of motion. Monkeys have much less mobility at the elbow because they are adapted more for speed than for swinging under branches. Apes use their large and powerful hands to grasp branches and to climb, whereas monkey hands tend to resemble their feet and are usually rotated so that the palm is on the ground or branch as they walk on all fours. Monkeys and apes both have relatively dexterous hands, but apes more so, and humans are the most dexterous of all.

Short legs, powerful hips. In the lower or hind limb, apes are perhaps a bit less distinctive, but we can still identify important differences from other primates. Apes all have shorter legs than arms, as I've already noted. The hips have a massive ball-and-socket joint, which both supports the body mass and also gives apes a great deal of mobility at the hip. This allows the larger apes to spread their body mass across several branches in the trees, reducing the chance that a branch will break under their weight.

In many of these features, humans are like other apes. Our lower backs are shorter than those of Old World monkeys but longer than those of the other great apes (we have five lumbar vertebrae on average, like gibbons). We have broad chests, shoulder blades positioned on the back, highly mobile shoulders, fully extendible elbows, wrists that rotate through a wide range of motion, and hands and feet that are quite distinct from each other in shape. To me this is clear evidence that humans evolved from a suspensory, apelike ancestor. To deny this, as, for example, the researchers working on *Ardipithecus* do, means that all of the similarities between ape and humans must have evolved independently. Given the huge number of traits we share with the apes, I find the hypothesis that they all evolved independently, and for no apparent reason, highly unlikely.

Of the many differences between apes and humans below the neck, almost all are related to one thing: our ability to walk on two feet. As we became bipedal, humans switched the ratio of arm-to-leg length. Longer legs made for more efficient walking and running. When compared with trunk length, our arms are much shorter than those of the apes. Our hands are smaller and less powerful than those of apes, except for our thumbs, which are very long and well supplied with strong muscles. Our hips have lost some of the mobility seen in apes, and our feet have become highly specialized for bipedalism. We have lost the opposable big toe found in all other primates, and our feet have become stable platforms with a variety of intricate mechanisms for making bipedalism efficient.

All of the features of human anatomy are built on the ape body plan. Without this evolutionary history we would not have our large brains and dexterous hands and our incredibly complex behavior. It is impossible to understand and explain the course of human evolution without recognizing the tremendous similarities between us and the apes.

WHEN DID APES FIRST ARISE?

Many researchers have tried to date when apes diverged from Old World monkeys by comparing DNA and calculating a rate of change based on the known rate of background mutation. This dating technique, known as the molecular clock, is the basis of many reconstructed branches on the tree of life, but it is not without its fair share of difficulties and controversies.

Here's how the molecular clock works. Mutations occur spontaneously in all genomes. It is an inherent property of DNA replication. This background mutation process is thought to have a characteristic rate, like the ticking of a clock. However, although we know that the rate can be different between organisms, it is likely to be similar among closely related animals like great apes. But how do we calculate the rate of mutation?

To calculate the rate, we need to know how much difference exists in the genomes of each species and how long it took to accumulate. Here is where paleontology and molecular biology join forces. If you know roughly how long ago one organism branched off from its closest relatives, and you know how many genetic differences there are between the two, you can calculate a rate. For example, if the fossil record tells us that species A and species B branched off 1 million years ago, and if the DNA tells us that there are one million differences in the genomes of each, then the rate of mutation is on average one per year. Once a rate is established, it can be used to estimate the divergence dates between other pairs of organisms. This process is called calibration.

In paleoanthropology, we commonly calibrate ape divergence times based on the divergence of the orangutan. Ancestors of modern orangutans are well represented in the fossil record. The oldest fossil members of the orangutan lineage are known from coal deposits in Thailand and from sand and clay deposits in Pakistan that are both about 12.5 million years old. We estimate that orangs diverged from the other great apes sometime before the first pongines (orangs and their ancestors) appear in the fossil record, possibly about 14 to 16 million years ago. We know how much genetic "distance" there is between orangutans and African apes and humans from molecular data. So, we can calculate an estimate of the rate of mutation among apes and humans from the amount of time since the divergence of orangs from African apes and humans and the number of genetic differences between the two groups. We then take this rate and use it to calculate a time of divergence between other pairs of species or groups of species. The time of divergence is the number of differences divided by the calibrated rate.

This type of analysis has been used to estimate when chimpanzees and humans diverged from a common ancestor. The estimates usually range from about 5 to 7 million years ago, although some estimates are as old as 14 million years and others as young as 3.5 million years. The 5-to-7-million-year estimate is broadly consistent with the fossil record of the earliest members of our evolutionary group, the hominins, although at 5 million, this would exclude a number of fossils widely accepted as hominins (*Orrorin*, *Ardipithecus*, *Sahelanthropus*) from our group. (We'll meet these fascinating recent finds in chapter 9.) Nevertheless, a broad if not quite complete consensus is emerging that humans and chimps diverged between 7 and 8 million years ago, gorillas around 9, and orangs, by definition, between 14 and 16 million years ago.

Of course, we are talking here about each individual lineage of living great apes and humans. What about, for example, the common ancestor of the living African apes and humans, an extinct species? When can that species, the first hominine, be expected to have arisen? Well, the answer is among the divergence dates of the living ape lineages. When we say that the orangutan, a pongine, branched off between 14 and 16 million years ago, we are essentially saying that the ancestor of the modern orangutan, the first pongine, branched off from the common ancestor of the African apes and humans, the first hominine, 14 to 16 million years ago. Since they are sister clades (most closely related organisms), hominines and pongines must have come into being at the same time, as a result of the same branching event.

Yet it is always wise to be a little cautious. The earliest members of any lineage will be very difficult to detect in the fossil record because they haven't changed much from their ancestors. Researchers need to estimate the amount of time that is missing, and it is really anyone's guess. As I said, most scientists accept a date that puts the branching-off of the pongines at about 14 to 16 million years ago, but there is no definitive evidence for this date. In other words, the oldest identifiable pongine fossils are 12.5 million years old, and we guestimate that another 1.5 to 3.5 million years are missing from the fossil record. In this particular case I don't think we could be too far off.

As we will see in later chapters, there were very interesting apes running around in Europe and Africa between 14 and 16 million years ago, and none of them looked very much like either pongines or hominines. They were more primitive looking, as we will see, and lacked characters such as limb structure and growth rates that are shared by pongines and hominines. So the split between pongines and hominines is not likely to have been before these more primitive apes lived. In addition, when we calculate a rate of change, we assume that the rate has remained constant through time and across different species. But this may not be the case.

Using this approach, molecular biologists have produced a number of different estimates for the time of divergence between the Old World monkeys and the apes. The different estimates reflect the fact that scientists have used different parts of the genome and made different assumptions about the rates of change. Dates for the Old World monkey–ape split range from about 31 to 38 million years ago. This means that we should start finding evidence of Old World monkeys that can be distinguished from apes in the fossil record around this time. Or does it?

While it is relatively easy to define living hominoids and to understand why humans are hominoids, when the fossil evidence is added to the equation it becomes much more difficult. This should come as no surprise. The attributes that characterize all hominoids today took time to evolve, and at the beginning of hominoid evolution, none of these features were present. There was a single population that was our last common ancestor with Old World monkeys.

Whatever the cause, a split happened. It probably occurred by chance, perhaps when two populations became separated from one another by a natural barrier, such as a river, or they may simply have drifted apart. Each population had its own unique combination of variations. With space between them and slightly different ecological settings, each was now subject to different selection pressures. Over time those would lead to different adaptations developing in each lineage that would eventually be pronounced enough that we can tell them apart in the fossil record.

However, because it took millions of years for even a few of the clear distinctions between monkeys and apes to evolve, it is a big challenge to decide which fossil is really an early ape. Even if the divergence of Old World monkeys and apes is as late as 31 million years ago, there is a big gap between this divergence estimate and the first appearance in the fossil record of anything we might want to call an ape or an Old World monkey. We take up this part of the story next.

CHAPTER 1

THE EARLY YEARS

Imagine that you are standing in a desert in northern Egypt at a place called the Fayum and are suddenly transported back 33 million years in the past. You are now standing in a forest rich in biodiversity. It's daytime, and up in the trees a creature about the size of a gibbon is moving cautiously about, using its forward-facing eyes with stereoscopic vision to detect the ripest fruit, its dexterous hands to pick and peel it, and its excellent sense of smell to make sure not to stray into the territory of another one of its kind. This is *Aegyptopithecus*. It is one of the few animals you see in Fayum that seems familiar to you, a primate that looks like a cross between a lemur and a monkey (figure 1.1).

Because Africa is separated at this time from Eurasia by a vast body of water, the Tethys Sea, many of the creatures of the Fayum are unique to Africa, and, because many have not left descendants, they are unfamiliar to us today. On the ground there are large, weird-looking animals vaguely resembling rhinos, but with horns that are side by side instead of one behind the other. You see a variety of browsing animals that mimic rhinos, pigs, deer, and other browsers but are not related to living forms either. There is a diversity of carnivorous animals (creodonts) that also went extinct without leaving descendants. You do see some elephants, but you do not recognize them because they are the size of cows and have no tusks. Since we are near the coast, you see mangroves and swamps, with crocodiles, turtles, and a variety of water birds. Because they spend most of their time in the water you are unlikely to spot a dorudon, an ancient whale with arms and legs!

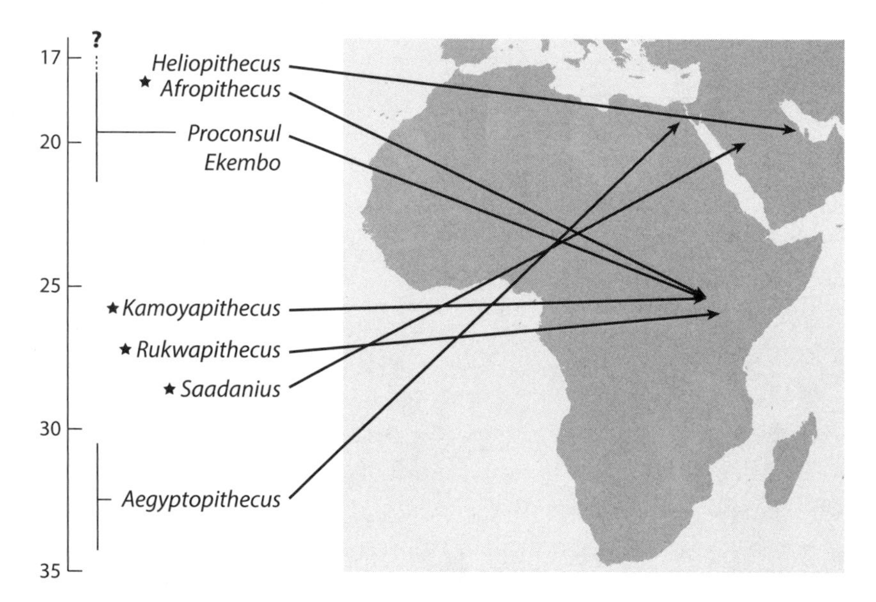

FIGURE 1.1. Timeline graphic showing time span from *Aegyptopithecus* through the early Miocene (ca. 33–17 Ma) and a map showing where the fossils were recovered (Egypt, *Aegyptopithecus*; Saudi Arabia, *Saadanius* and *Heliopithecus*; Kenya, *Proconsul*, *Ekembo*, and *Kamoyapithecus*; Tanzania, *Rukwapithecus*).

In the trees with *Aegyptopithecus* there are many other primates, including other catarrhines, more primitive anthropoids, and even tarsiers. There is also an array of insectivores, bats and shrewlike animals and rodents, the latter mostly ground dwelling. While the composition of the fauna at Fayum is unlike that of any modern forest, because most of the species are extinct, together they comprise an ecosystem much like the relatively dry forests of parts of equatorial Africa today.

Although *Aegyptopithecus* lived before the Old World monkeys and apes diverged from one another, some scientists think that the common ancestor of Old World monkeys and apes may have resembled *Aegyptopithecus* (see plate 4). We are blessed with a rich fossil record of *Aegyptopithecus* and other early catarrhines from the bountiful deposits of the Fayum depression in Egypt. Researchers first discovered fossil primates at Fayum at the beginning of the twentieth century, and researchers are still working there. For much of the last half of the twentieth century, a luminary in the field, Elwyn Simons, directed scientific research at the Fayum. Simons was the dissertation supervisor of many of the most accomplished researchers in paleoanthropology, and I am happy to know him. When we first met I learned that he had supported my application for a grant to have another look at the fossil apes of Europe.

Compared with most apes, *Aegyptopithecus* was small. Based on fossil evidence, scientists think it weighed about 6 kilograms (about 16 lb.). It had a large face with a prominent snout that had to be large in order to house well-developed organs, thus providing *Aegyptopithecus* with an excellent sense of smell, a feature it shares with living prosimians. In contrast, apes have a much flatter profile. Most prosimians mark their territories with oily secretions from scent glands and have a better-developed sense of smell than most monkeys and apes, which do not mark territories with smelly secretions. In this way, prosimians are more like many other animals that rely on scent, whereas apes have shifted the dominant sense from the nose to the eyes. We also see this reflected in the brain of *Aegyptopithecus*, which had comparatively large lobes in its brain devoted to processing smell. The brain of *Aegyptopithecus* was small and, when scaled relative to body mass, was closer to modern prosimians than to most monkeys and apes. Of course, 33 million years ago, the brains of the ancestors of living prosimians had not reached their current size, so for their time period, *Aegyptopithecus* was probably fairly brainy. In many ways *Aegyptopithecus* was an intermediate form between prosimians and anthropoids, even though *Aegyptopithecus* was not only an anthropoid but also a catarrhine.

Remember when I said above that the first members of a group will be hard to discern from their closest relatives because most of the characteristics that allow us to classify animals in a particular group had not yet evolved? This is the case with *Aegyptopithecus*. Luckily for us, however, the teeth give *Aegyptopithecus* away. Like all catarrhines, including us, *Aegyptopithecus* had a dental formula of 2:1:2:3. Let me explain. We have 32 teeth in our mouths, if they all come in normally (many modern humans have impacted or even missing last molars, or wisdom teeth). Each of our jaws—the upper jaw, or maxilla, and the lower jaw, or mandible—has 16 teeth, 8 on each side of each jaw. Eight times four is 32. These 8 teeth are used to identify the dental formula. In humans, they consist of two incisors, one canine, two premolars and three molars. New World monkeys, for example, have three premolars instead of two, as do most prosimians. So *Aegyptopithecus* shares with catarrhines the 2:1:2:3 dental formula.

In addition to the number of teeth in its mouth, *Aegyptopithecus* shares something else with most other catarrhines: the form of its teeth, particularly the molars. *Aegyptopithecus* has hominoid-shaped molars. In fact, it and other close relatives from the Fayum are sometimes referred to as dental apes. The upper molars have four cusps with three arranged in a triangle (trigon) and the fourth at the back inner corner of the tooth. The lower teeth are arranged in a distinctive pattern of five cusps, with a Y-shaped groove separating them (figure 1.2). We will run into this so-called Y-5 cusp pattern again later in this book, but not among Old World monkeys, which have their own specialized dental morphology.

For many years *Aegyptopithecus* was interpreted as a fossil ape precisely because of its molar anatomy. Its broad, flat molars and Y-5 dental pattern were considered to be shared uniquely with apes. We now know that this molar morphology was shared by the ancestors of Old World monkeys as well, but, as I mentioned, the monkeys have moved on. In a very real sense, they are more evolved than we are, at least in molar morphology.

In the Old World monkeys, tooth shape kept evolving as their teeth became ever more specialized for processing leaves and other fibrous foods. We call their teeth bilophodont, that is, a tooth (dont) with two (bi) cusps or bumps on the tooth connected by a ridge (lophs). Among anthropoids, only Old World monkeys have bilophodont molars, which they use like scissors to slice through vegetation. They are bilophodont in both their upper and lower molars, which therefore look similar to each other, unlike the easily distinguished upper and lower molars of *Aegyptopithecus* and the apes.

The oldest known Old World monkey fossils (*Prohylobates* and *Victoriapithecus*) have bilophodont molars but with traces of the

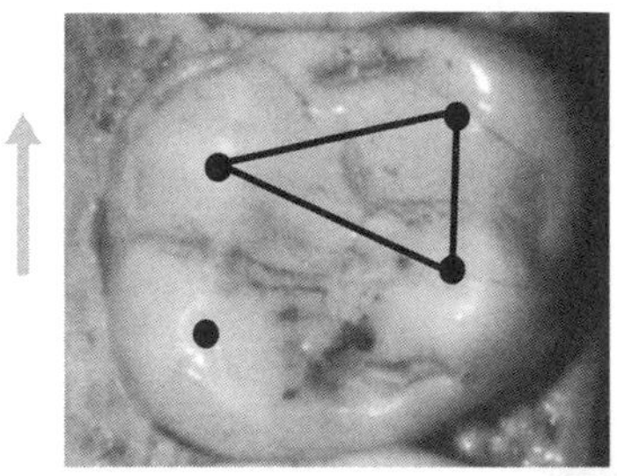
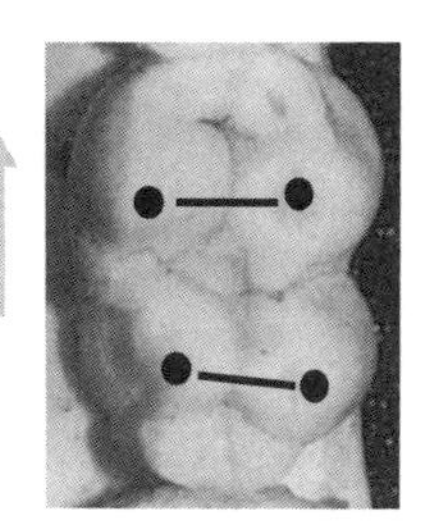
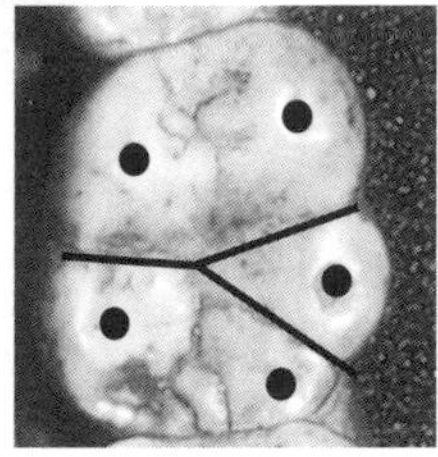

FIGURE 1.2. Comparison of the upper and lower molar teeth of Old World monkeys and apes. Old World monkeys have bilophodont molars with a sharp crest joining the front and back cusps. The upper and lower molars closely resemble each other. In apes the upper molars have four cusps, three of which are arranged in a triangle with the fourth cusp attached outside the long edge of the triangle. The lower molars typically have five cusps arranged in the Y-5 pattern as described in the text. Arrows point to the front of the mouth. (Images by author.)

crests that are found in the teeth of hominoids and *Aegyptopithecus*. While we cannot say with certainty what additional attributes *Prohylobates* and *Victoriapithecus* had to aid in their leaf-eating habits, living Old World monkeys have specializations in their cheeks and stomachs to help them get enough nourishment from their mostly leaf-based diets.

Paleontologists think that *Aegyptopithecus*, like nearly all monkeys and all apes, was active during the day, or diurnal. We conclude this because *Aegyptopithecus* has eye sockets similar in size to those of living diurnal monkeys with skulls of roughly the same size. Nocturnal primates always have large eyes, in order to capture more of the available light at night. *Aegyptopithecus* is also sexually dimorphic in the size and shape of its canines, which means, simply, that the canines of males and females look different. Males have elongated bladelike canine teeth, whereas females have shorter, conical canines. This is typical of anthropoids that live in social groups in which males compete for rank and mates. *Aegyptopithecus* had relatively stocky arms and legs compared with most monkeys, and powerful hands and feet for grasping. It was probably a relatively cautious climber, unlike most monkeys living today, which move with great speed and agility in the trees. *Aegyptopithecus* is the best approximation we currently have of the animal from which Old World Monkeys and apes evolved.

Aegyptopithecus was a not-quite-ape. But what did the first true ape look like? We lack fossils that represent the earliest populations of Old World monkeys and apes. Again, this is not surprising and should not be taken as an indication that the fossil record is so inadequate that we should give up trying to interpret it. It is incomplete, to be sure. In fact, the fossil record can be described as consisting of a few words per chapter of a book thousands of chapters in length. Try putting together the whole story from that! You will never succeed, but you will learn a great deal by looking at the fossils that we do have.

For example, although we have many fossils from the skull, teeth, and limbs of *Aegyptopithecus,* we do not know from the fossil evidence whether or not it had a tail. I think it probably did. All Old World monkeys have tails, even if in some cases you can't see them under the fur. All apes lack tails. Given that *Aegyptopithecus* lived before the apes and Old World monkeys evolved and that it gives us a good idea of what the common ancestor of Old World monkeys and apes looked like, if *Aegyptopithecus* did not have a tail, then monkeys would have had to re-evolve one. No one I know thinks that this is even remotely likely. So, despite the ape-looking teeth of *Aegyptopithecus,* we know that more change had to come before we can say we are dealing with true hominoids. *Aegyptopithecus* was, broadly speaking, a common ancestor of Old World monkeys and apes, but not the last common ancestor. There is a gap in the fossil evidence between *Aegyptopithecus* and the populations that diverged into the first members of the Old World monkey and ape clades. Recently though, a new taxon has been discovered that partly fills this gap and comes from what seems at first to be an unlikely place.

Saadanius is a fossil about 28 million years old, occupying the time gap between *Aegyptopithecus* and the first potential apes that I will discuss below. It is from Saudi Arabia, which may seem odd, but in fact at the time Saudi Arabia was part of mainland eastern North Africa, and its ecology was similar to that at Fayum. The *Saadanius* site is actually about halfway between the Fayum to the north and the first potential fossil-apes sites in Kenya and Tanzania. *Saadanius* is larger than *Aegyptopithecus*, as are the oldest hominoids, but is otherwise quite similar in many anatomical details to *Aegyptopithecus*, with one important exception. At the base of the skull in all modern and most fossil catarrhines, there is a tube, the ectotympanic tube, running from the outer to the middle ear. All hominoids and cercopithecoids, living and extinct, have this bony tube on the bottoms of their skulls. *Aegyptopithecus* does not, which is another indication that *Aegyptopithecus* is more primitive than the last common ancestor of living catarrhines. On the other hand, *Saadanius* has the tube; so, it is more evolved and closer to the last common ancestor of apes and Old World monkeys, because it shares a feature not found in *Aegyptopithecus* or any non-catarrhine primate. Since *Saadanius* is only 28 million years old, it is probably younger than the split between cercopithecoids and hominoids, which is dated by the molecular clock to be at least 30 million years ago. It was mostly likely a survivor, a relic, of a lineage that existed earlier. Even so, it gives us a lot of help in getting closer to that elusive last common ancestor. By about 26 million years ago, we start to pick up evidence of another catarrhine, one that we might just be able to call an ape.

KAMOYAPITHECUS: THE FIRST APE?

The oldest fossil that represents an ape is probably a piece of an upper jaw representing the genus *Kamoyapithecus*, which is named in honor of the legendary Kenyan fossil hunter, Kamoya Kimeu.

Kamoyapithecus is known from the site of Lothidok (or Losodok), in northern Kenya, and dates to about 26 million years ago, which falls in the geological epoch known as the Oligocene ("epoch" is the formal term for the main subdivisions of periods; the Oligocene and the Miocene, which comes right after the Oligocene, are epochs in the Tertiary period) *Aegyptopithecus* is also from the Oligocene but is older, at about 31 to 33 million years. Lothidok, like Fayum, was a forest full of archaic African mammals, since the sea barrier to faunal exchange with Eurasia was still in place. It was probably ecologically similar to Fayum, a bit on the dry side compared with many modern tropical forests, and somewhat drier than the forests in which our next characters in this story, *Proconsul* and *Ekembo*, lived.

Why do I think that *Kamoyapithecus* is an ape? It has teeth that, although larger, resemble those of *Aegyptopithecus*. There are subtle differences in the degree of development of certain attributes of the molars that potentially indicate an evolutionary change between *Aegyptopithecus* and *Kamoyapithecus*. In addition, the canines of *Kamoyapithecus* are massive, unlike the more slender male canines of *Aegyptopithecus* and more like those of *Ekembo* (see below). However, given the differences in size between the two, it is hard to say if the dental differences are just size related or are evolutionarily significant. The anatomy of the teeth suggests that *Kamoyapithecus* had an apelike diet consisting mainly of fruits, but then so did *Aegyptopithecus*. Frankly, if it had been found alongside *Aegyptopithecus*, it would probably be considered to be a larger version of the same animal.

There is so little to work with of *Kamoyapithecus*, which is known only from a few pieces of the upper jaw and some isolated upper teeth. In 2013 we caught a bit of a break with the publication of a new ape fossil from Tanzania. This creature, called *Rukwapithecus*, is about the same age as *Kamoyapithecus*. But as luck would have it, *Rukwapithecus* is known only from a lower jaw, so the two cannot be compared directly. However, on the basis of that jaw, *Rukwapithecus* is much smaller than *Kamoyapithecus* and looks distinctly more primitive. While Nancy Stevens (who very kindly showed me a beautiful cast recently) and colleagues think that *Rukwapithecus* is an early hominoid, I think it more closely resembles *Pliopithecus*, an ancient offshoot of the catarrhines (I will discuss *Pliopithecus* a bit later on). Either way, it is amazing to have a fossil like this in Tanzania, and it tells us that we should be prepared for any possibility in paleontology.

Kamoyapithecus and *Rukwapithecus* are known from the right time period and in the right place, East Africa near the end of the Oligocene, to be viable representatives of the ancestors of later apes. It would be nice to have some limb bones, to see if there are differences in the direction of later apes, and more bits of the skull, to see how they compare with *Aegyptopithecus* and *Saadanius*, but for these we will have to wait for future discoveries. But in sediments that are just a few million years younger, researchers have found fossils with a more definitive ape look.

ENTER *PROCONSUL* AND *EKEMBO*

Proconsul is an excellent example of an intermediate taxon between the more primitive monkey-like primates and more modern apes. A great diversity of fossil apes flourished between about 20 and 17 million years ago, and one of them, although we do not know exactly which one, is probably more closely related to living apes than the others are. Together they represent the ancestors of all living apes, including humans (table 1.1).

Of these early apes, we know *Proconsul* and its close relative *Ekembo* best. The oldest *Proconsul* fossil discovered thus far is from a site in western Kenya called Meswa Bridge. Unfortunately, this early sample is limited and composed mostly of juveniles, which complicates any comparison with adult fossils of *Proconsul* uncovered elsewhere. My colleagues Terry Harrison and Peter Andrews have recently named a new species based on the Meswa Bridge fossils, *Proconsul meswae*. *Proconsul* is better known from somewhat younger deposits elsewhere in Kenya.

As an aside, Terry and I have known each other since we were graduate students, and I am not sure that we have ever agreed on anything. Nevertheless, I admire his work and I like him as a person. He is among the hardest working paleoanthropologists I know. Peter Andrews is the person who, along with David Pilbeam (see chapter

TABLE 1.1. Miocene “Apes” (Non-cercopithecoid Catarrhhines).

	AGE[1]	MA	GENERA	IMPORTANT LOCALITIES	COUNTRY	MATERIAL[2]
1	Oligo.	28–29	*Saadanius*	Harrat Al Ujayfa	Saudi Arabia	Facial cranium
2	Oligo.	25.2	*Rukwapithecus*	Rukwa Rift	Tanzania	Mandibular fragment
3	Oligo.	25	*Kamoyapithecus*[3]	Lothidok	Kenya	Craniodental fragments
4	e M	21	”Proconsul meswae”[4]	Meswa Bridge	Kenya	Craniodental fragments
5	e M	?20–17.5	*Morotopithecus*	Moroto	Uganda	Cranial, dental, postcrania
6	e M	?20–17.5	*Kogolepithecus*[3]	Moroto	Uganda	Dental
7	e M	19	*Ugandapithecus*[4]	Napak/Songhor	Kenya/Uganda	Cranial, dental, postcrania
8	e M	19	*Xenopithecus*[3]	Koru	Kenya	Craniodental fragments
9	e M	19	*Proconsul*	Songhor/Koru	Kenya	Cranial, dental, postcrania
10	e M	19	*Limnopithecus*[3]	Koru/Songhor	Kenya	Craniodental
11	e M	19	*Rangwapithecus*[3]	Songhor	Kenya	Craniodental
12	e M	19	*Micropithecus*[3]	Napak/Koru	Kenya/Uganda	Craniodental
13	e M	19	*Kalepithecus*[3]	Songhor/Koru	Kenya	Craniodental fragments
14	e M	18–19	*Dendropithecus*[3]	Rusinga/Songhor/Napak/ Koru	Kenya/Uganda	Cranial, dental, postcrania
15	e M	19	*Lomorupithecus*	Napak	Uganda	Craniodental fragments
16	e M	19.5–17	*Ekembo*	Rusinga/Mfangano	Kenya	(Cranial, dental, postcrania)+
17	e M	17.5	*Turkanapithecus*	Kalodirr	Kenya	Cranial, dental, postcrania
18	e M	17.5	*Afropithecus*	Kalodirr	Kenya	Cranial, dental, postcrania
19	e–m M	17.5–15	*Simiolus*	Kalodirr/Maboko	Kenya	Cranial, dental, postcrania
20	e–m M	17.5–15	*Nyanzapithecus*	Rusinga/Maboko	Kenya	Craniodental fragments

21	e					Dental
22	e					
23	e					
24						
25						
26						
27						
28						
29						
30						
31						
32						
33						
34	m–l M	?13.5–				
35	l M	10	*Rudapithecus*	Rudabánya, Alsótelekes		
36	l M	10–9	*Hispanopithecus*	Can Llobateres/Can Ponsic	Spain	
37	l M	11–8?	*Neopithecus*[6]	Salmendingen	Germany	Third molar
38	l M	10	*Ankarapithecus*	Sinap	Turkey	Cranial, dental, postcrania
39	l M	9.5	*Samburupithecus*	Samburu	Kenya	Craniodental fragments
40	l M	10.25	*Chororapithecus*	Chorora Fm	Ethiopia	Isolated teeth
41	l M	9.85	*Nakalipithecus*	Nakali	Kenya	Mandible and teeth
42	l M	9.5	*Ouranopithecus*	Ravin de la Pluie	Greece	(Craniodental)+, 2 phalanges

(*continued*)

TABLE 1.1. *(Continued)*

	AGE[1]	MA	GENERA	IMPORTANT LOCALITIES	COUNTRY	MATERIAL[2]
43	l M	9–8	*Graecopithecus*	Pygros	Greece	Mandible
44	l M	8–6	*Lufengpithecus*	Lufeng, Yuanmou, Shuitangba	China (Yunnan)	(Cranial, dental)+, postcrania
45	l M	8–7	New taxon	Çorakyerler	Turkey	Mandible, maxilla
46	l M	7	*Oreopithecus*	Baccinello/Monte Bamboli	Italy	(Cranial, dental, postcrania)+
47	l M	7–6	*Sahelanthropus*	Toros-Menalla	Chad	Craniodental
48	l M	6.5	*Indopithecus*	Haritalyangar (Siwaliks)	India	Mandible
49	l M	6	*Orrorin*	Lukeino	Kenya	Craniodental, postcrania
50	l M	5.8–5.2	"*Ardipithecus*"[7]	Alayla (Middle Awash)	Ethiopia	Craniodental, postcrania
51	Pl.	1–0.3	*Gigantopithecus*	Liucheng	East Asia	Mandibles, teeth

NOTE: These fifty genera include taxa from the Oligocene and Early Miocene that share mainly primitive characters with the Hominoidea but that appear to be derived relative to Pliopithecoids and Propliopithecoids. See text for further discussion.

[1] Oligo. = Oligocene; e M = early Miocene; e–m M = early–middle Miocene, m M = middle Miocene, m–l M = middle–late Miocene; l M = late Miocene; Pl. = Pleistocene.

[2] Material briefly described by part representation. Dental = mainly isolated teeth; dental, mandible, maxilla = known only from these parts; craniodental fragments = teeth and few cranial fragments; craniodental = larger samples of more informative cranial material; cranial, dental, postcrania = good samples from each region; ( )+ = very good representation of parts in parentheses.

[3] Unclear attribution to Hominoidea.

[4] Large taxon possibly distinct from *Proconsul*.

[5] Middle Miocene samples attributed to this taxon are more fragmentary and may not be congeneric.

[6] The *Neopithecus brancoi* type is an isolated third molar that some attribute to the Pliopithecoidea while others to *Dryopithecus* (Begun and Kordos, 1993). I regard it as a nomen dubium.

[7] "*Ardipithecus kadabba*." The type of the genus, *Ardipithecus ramidus*, is younger and shares derived characters with australopithecines not present in *A. kadabba*, making *Ardipithecus* paraphyletic, in my opinion.

7), redirected interest in paleoanthropology from a more or less exclusive focus on hominins to giving some attention to fossil apes. I met Peter and Terry in 1980 on my first trip to visit European museums. After a pint with Terry, I returned to Peter's lab to continue working on the *Proconsul* fossils in his care, and I blurted out something I thought Terry had told me at lunch, which was that he was about to completely revise *Proconsul*, which Peter had done just a few years ago. Who says that? Peter, very British, was unperturbed, but I could tell almost immediately that I had inserted my foot very deeply into my mouth. As it turns out, Peter Andrews' ideas about early Miocene apes are mostly intact 35 years later, with a few tweaks.

Proconsul was first identified as a candidate for the ancestry of apes and humans in 1933. The species name *Proconsul africanus* was inspired by Consul, a very popular performing chimpanzee. Consul was dressed in a suit, used a cane, drank liquor from a glass, and smoked cigars, which was immensely entertaining to turn-of-the-nineteenth-century English crowds. As a child, Arthur Tindell Hopwood, who named the taxon, was entertained by Consul and apparently never forgot him. Hopwood recognized the similarities between the fossils that he had in front of him and chimpanzees, and he promoted the idea that *Proconsul* is the ancestor of living chimpanzees, and thus our ancestor as well.

Proconsul still occupies this enviable position. Today we would be careful to say that *Proconsul* is what is known as a stem hominoid. Similar to the way the trunk of a tree grows before the branches, stem taxa appeared before the branching of modern taxa from their common ancestor. In this case, *Proconsul* is a stem hominoid because it evolved before any of the living hominoids or their fossil relatives branched off from one another. Crown taxa are those that evolved after the branching had begun. So we would say that *Proconsul* is a crown catarrhine, because it evolved after apes and Old World monkeys had diverged; that is, it is on the hominoid branch of the catarrhine tree, not the trunk. It is likely that a population derived from a species of *Proconsul*, or a close relative, is ancestral to all living apes (figure 1.3).

The best-known collection of early Miocene apes comes from the site of Rusinga Island, Lake Victoria, Kenya. Known as *Proconsul* for

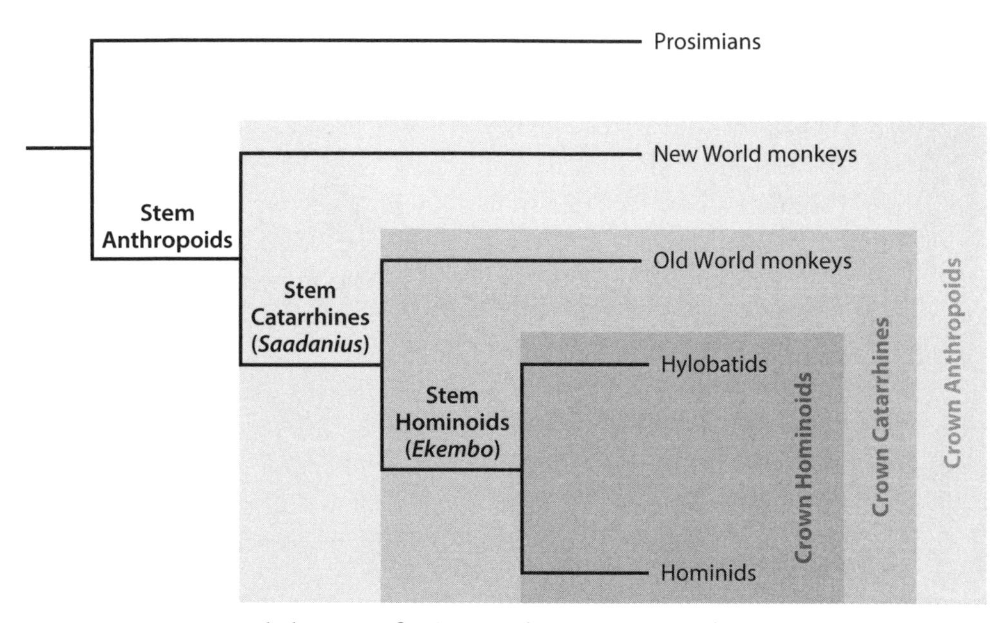

FIGURE 1.3. Cladogram of primates showing stem and crown taxa. Stem taxa include extinct lineages, and crown taxa include all living members and their fossil relatives. The shaded boxes delineate crown groups. Stems lead to their respective crown groups. For example, *Saadanius* is a crown anthropoid (*lightest box*) and a stem catarrhine (leading to crown catarrhines). *Saadanius* is more closely related to catarrhines than to non-catarrhine anthropoids, such as New World monkeys. *Ekembo* is a crown catarrhine (*darker box*) and a stem hominoid because it is more closely related to living hominoids than to any other catarrhine, but it is not directly related to any individual group of living hominoids.

many years, the specimens from Rusinga were recently renamed *Ekembo* by Kieran McNulty, Jay Kelley, and I. *Proconsul* and *Ekembo* are closely related. Recognition of the new genus reflects the diversity of apes in the early Miocene. So from now on I will be talking about *Ekembo*, which is known from many more specimens than *Proconsul.*

Rusinga is one of the most continuously worked sites in the world, having been surveyed and excavated on and off from the 1920s to the present. We know more about the biology and behavior of *Ekembo* from work carried out on the fossils from Rusinga than we know about most other fossil primates. Rusinga is a fascinating site. It preserves detailed information about the environment in which the apes lived that is rare for a fossil locality, due in part to the nature of the sediments. Today Rusinga is on the shores of Lake Victoria, but at the time that *Ekembo* and other primates lived, the lake had not formed. Instead, the landscape was forested, with several volcanoes. These volcanoes were unusual in that they ejected what geologists call carbonatite tuffs, volcanic ash with a hyperalkaline (high pH) chemical composition that makes the ash relatively cool and preserves fossils extremely well. The ash beds at Rusinga have yielded fossil grasshoppers and grubs of various insects, as well as the bodies of chameleons, preserved as if they had died at Pompeii. It is here that scientists have recovered the best collection of Miocene primates from any site in Africa.

A biologist walking through the forests at Rusinga would have seen a woodland with a richly diverse flora and fauna but a woods not as densely forested as a modern tropical forest. It has been described as a dry, seasonal, equatorial forest, perhaps most like the forests today in Central America and parts of equatorial Africa. At Rusinga, many of the trees lost their leaves in the fall, and there were dry and wet seasons. However, the climate was still quite pleasant and equable, with less temperature seasonality than we will see later on in the habitats of Miocene apes. Like all dry forests, those at Rusinga were more heterogeneous than tropical rain forests and had, for example, microecological settings ranging from more open country with widely spaced trees to wetter, even marshy, conditions. Many of the animals that lived in Rusinga could probably have survived in several of these microzones, including *Ekembo*, but the scientific consensus is that *Ekembo* was a forest dweller.

A biologist studying the fauna of Rusinga would have taken great pains to spot the primates living there and to document their ecological and behavioral preferences. She would have seen a variety of basically monkey-like primates that are fairly agile in the trees, including *Dendropithecus*, *Limnopithecus*, *Nyanzapithecus*, and *Ekembo*. While agile and arboreal, these primates moved in the trees much like living monkeys, walking on the tops of branches. It is not clear how *Dendropithecus*, *Limnopithecus*, and *Nyanzapithecus* are related to living apes or even to *Ekembo. Limnopithecus* was a small catarrhine, maybe three or four kilograms (about 6–8 lb.) in size, whereas *Dendropithecus* was somewhat larger.

Dendropithecus is known from several limb bones that are long and slender, suggesting that it could move with agility in the trees. Its molars are "crestier" than those of *Ekembo*, suggesting that it may have been more specialized in its diet, perhaps processing more fibrous foods. *Nyanzapithecus* has unusual teeth with tall, pointy cusps that make it difficult to assign it to a branch on the tree of life. We're just not sure who its closest ape relatives are.

At night our biologist might have spotted a *Progalago* or a *Komba*, relatives of living galagos (bush babies), high in the canopy, or *Mioeuoticus*, a loris-like animal, clinging to a small tree trunk. Rusinga would have been a favorite place for primatologists, with its three species of *Komba* and two species of *Ekembo*, as well as the primates I mentioned above. Besides the primates, biologists would have been documenting a huge diversity of small mammals, including dozens of rodents, insect eaters (formerly known as Insectivora but now split into a number of orders), bats, and rabbits. They would also have found some holdovers from the Oligocene, including hyracoids—weird, rotund animals that look like groundhogs. They are actually related to elephants, anteaters, and other endemic African taxa as well as to hyraxes and manatees. Hyracoids were also present in Fayum and Lothidok but were more primitive and diverse. One species from the Fayum was the size of a small rhino. Primitive relatives of pigs, hippos, and elephants lived in the vicinity, as did a large number of carnivores. Early rhinos and ruminants (animals that chew their cud) were also present, including a predominance of species that probably had a preference for forests. Biologists would have seen several types of forest-living tragulids (mouse deer, or chevrotains) of various sizes, representing a diversity that was typical of the Miocene throughout the Old World, whereas today the mouse deer is largely confined to the forests of Southern Asia with one species left in Africa. Finally, our biologist studying the fauna of the Rusinga area would have been captivated by a number of animals that have no direct descendants today, including anthracotheres (most likely related to hippos) and creodonts (an extinct group of carnivorous animals distantly related to modern carnivores). Rusinga was also home to chalicotheres, a very strange animal that walked on its knuckles like a gorilla. They had long forelimbs and short legs and are related to rhinos and horses.

She also would have spotted the occasional deinothere, a distant relative of elephants with downward-curved tusks. Rusinga was in fact an amazingly diverse ecological setting. We know a little less about the plants and insects from Rusinga, but all evidence also points to a rich, dense forest with many species of trees and shrubs and few grasses and sedges. It was probably ideal habitat for animals that live primarily in the trees or off products of the trees.

Ekembo is one of the best-known fossil primates. Many specimens have been recovered from a large number of sites in both Kenya and Uganda, but the most complete and most numerous fossils come from Rusinga and the nearby site of Mwfangano. At one especially rich site on Rusinga Island, the Kaswanga Primate Site (KPS), scientists have recovered the partial skeletons of nine individuals. I worked on the phalanges (finger and toe bones) from KPS, about 300 in number, and published what one of my own coauthors called the longest and most boring article ever. Actually, I laugh every time I think of it. It is long and boring, but full of data.

I did learn a lot about the extremities of *Ekembo*. I can't begin to stress how rare and valuable the discovery of even a single partial skeleton is for understanding the biology and evolutionary history of a fossil taxon. The vast majority of fossil vertebrate taxa are known only from jaws and teeth (this even includes some fossil hominins). If limb bones are not found right next to jaws and teeth, it is often very difficult to know to which animal they belong. This can lead to a confusing and sometimes counterproductive situation. This dilemma is illustrated nicely by the curious case of *Paidopithex*.

Paidopithex rhenanus is a species name based on a beautifully preserved femur or thigh bone with a colorful history. The bone was discovered early in the nineteenth century at a site called Eppelsheim, near Mainz, in Germany, and was sent to luminaries in the nascent field of vertebrate paleontology, including the great founding father of the field, Georges Cuvier, who had established that species such as giant sloths and mastodons had once roamed the Earth. He was instrumental in making vertebrate paleontology a science and advocated the careful reconstruction of fossil taxa and a detailed understanding of their comparative anatomy. Based on anatomical and geological criteria, Cuvier also famously debunked pseudofossils, showing that most of these pseudofossils were either very poorly reconstructed or that they were not fossils at all, given their geological context or the other bones with which they were associated.

Although Cuvier made the great breakthrough in recognizing that such behemoths as mastodons once lived in the Paris Basin, he also believed that periodic floods like the one recounted in the Bible had consigned these past forms of life to local extinction. He was not convinced that they were indeed extinct and held that many may still live in far off places. Why the contrast between rigorous dispassionate analysis of fossils and their context at a given site and an apparently irrational belief in recurring floods replacing all existing life in a region with new forms?

Cuvier was a devoutly religious man, and he could not accept the idea that forms of life created by God could disappear completely from the face of the Earth. He also could not bring himself to accept the idea of transformationism (evolution) and was a brutal critic of the works of Jean-Baptiste Lamarck. He especially disapproved of the idea that there could be fossil humans and that humans could have evolved from other primates. Cuvier famously exclaimed, "L'homme fossile n'existe pas" (fossil man does not exist). So when the femur was sent to him, he did not comment on it. History does not record his motives for ignoring this otherwise beautiful specimen, but I suspect that with his outstanding anatomical background, Cuvier saw the primate features of the femur and was conflicted. The femur received scant attention until the end of the nineteenth century, when it was rediscovered and given the name *Paidopithex*.

The *Paidopithex* femur is long and straight and has a small hip joint, unlike the femora of living apes. Some early researchers claimed that it was the femur of a little girl, though this idea was more informed by the belief that there were no fossil apes and by Cuvier's influence than by anything in the anatomy of the specimen. Others saw rightly that it was a bone from a nonhuman primate and, based on its size, surmised that it must have come from

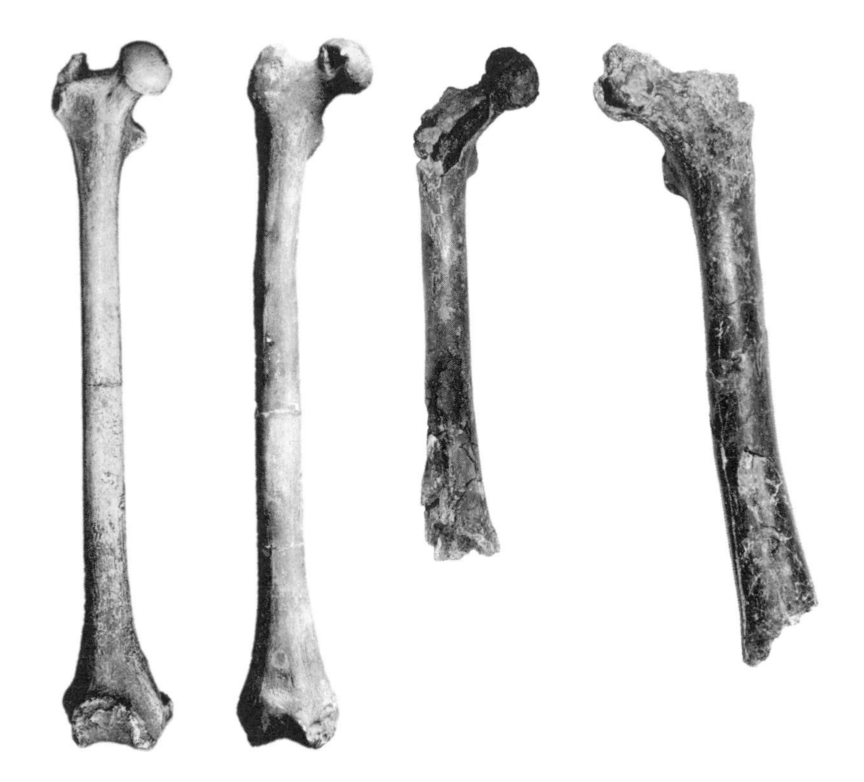

FIGURE 1.4. Femora of *Epipliopithecus*, *Paidopithex*, and two from *Rudapithecus*. The femur of *Epipliopithecus* (*far left*) is much smaller than that of *Paidopithex* (*second from left*), but when scaled to the same length, they look very similar and quite unlike that of a hominoid. The *Rudapithecus* femora (*right* and *far right*) strongly resemble those of living great apes, having much larger heads and much thicker, shorter shafts. (Images by author.)

an animal about the same size as *Dryopithecus*, an early great ape (see chapter 7), so it was widely attributed to this species. This is a common practice in paleontology. If researchers feel that two separately named organisms are really the same thing, they use the first name proposed and relegate the second name to a synonym of the first. So *Paidopithex rhenanus* became *Dryopithecus rhenanus*. Unfortunately, no *Dryopithecus* teeth are known from Eppelsheim, and no femora were known, at least until recently, for *Dryopithecus* (figure 1.4).

In 1960 a spectacular discovery in Slovakia of three partial skeletons of small monkey-like primates called *Epipliopithecus* was published in an excellent monograph by one of the luminaries of twentieth-century European vertebrate paleontology, Helmut Zapfe. I had the great pleasure of meeting Professor Zapfe twice in Vienna, where he worked in the Natural History Museum. Professor Zapfe died a few years ago, but his influence on European vertebrate paleontology will be everlasting.

The specimens of *Epipliopithecus* are so well preserved because they were found in the remains of a sinkhole just within the city limits of Bratislava. A sinkhole, or karst, is a geological formation that forms when limestone is dissolved by underground water, which creates caverns (the kind with stalagmite and stalactites). When the underground chamber gets large enough to approach the surface, the roof collapses, creating a sinkhole. You hear occasionally about buildings or cars being swallowed up by sinkholes. In the case of these fossils, they were probably washed into an open sinkhole after heavy rains. Many other well-preserved remains of animals that lived with *Epipliopithecus* were also found in the sinkhole.

The *Epipliopithecus* sample includes several femora, which were not previously known for the group to which *Epipliopithecus* belongs, the Pliopithecoidea. Pliopithecoids include the first described primate genus, *Pliopithecus*, which was named in 1836 by one of Cuvier's more illustrious students, Edouard Lartet. Cuvier himself had actually named a primate earlier, the prosimian *Adapis*, but he thought it was an artiodactyl (even-toed ungulate). Pliopithecoids were widespread across Eurasia and lived at roughly the same time as the hominoids that we will learn about later. But they were much more primitive than hominoids. For example, they lacked a completely formed ectotympanic tube, the ear structure that distinguishes all modern catarrhines from *Aegyptopithecus* and non-catarrhines. Pliopithecoids diverged from the rest of the catarrhines some time after *Aegyptopithecus* and before the Old World monkey–ape split. Their origin is a mystery because, while they must be over 30 million years old, the only fossils we have for them are no older than 18 million years. But once again, I digress.

Pliopithecoids, in addition to being very primitive, were smaller in general than great apes, and the *Epipliopithecus* femora are much smaller than the femur from Eppelsheim, so direct comparisons were not made between the two until the mid-1980s.

At that time, I was beginning my dissertation research on the fossil apes of Europe, and I began to study both the great apes and the pliopithecoids. I noticed a strong similarity between another specimen collected very early in the nineteenth century—a tooth from Salmendingen, near Stuttgart—and the relatively large pliopithecoid from Rudabánya in Hungary, *Anapithecus* (chapter 8). So I realized that there were larger pliopithecoids than *Epipliopithecus* and that some of them lived in Germany, very close to Eppelsheim. This made me wonder if the famous femur from Eppelsheim was a pliopithecoid and not *Dryopithecus* (a great ape), especially since the Eppelsheim femur is very different from an ape femur. My comparisons to Zapfe's *Epipliopithecus* femora led me to conclude that they were both from the same kind of animal, a quadruped that was good at climbing trees but walked on the tops of branches rather than swinging below them. So the Eppelsheim femur went from being a little modern human girl to *Paidopithex* and then to *Dryopithecus*, a great ape, and then back to *Paidopithex*, but now as a pliopithecoid. All of this because it was found in isolation, without nearby jaws or teeth.

In the intervening twenty-five-plus years, femora from three European fossil great apes have been found, and in all three cases they are different from the specimens from Eppelsheim and from *Epipliopithecus*, which supports the idea that *Paidopithex* is a large pliopithecoid and not a great ape. This matters to our understanding of the biology of our ancestors, because the way in which the animal represented by the Eppelsheim femur moved around in the trees was very different from the way we now know that extinct European great apes moved about. If we only had the Eppelsheim femur to go on, which was the case until recently, and we continued to think it was a hominoid, then we would have had a very wrong impression of how an animal very close to our evolutionary lineage moved and got around.

The point of this long aside is to show that when limb bones aren't found with jaws and teeth, it is *very* difficult to know how to assign the isolated limb bones, and that basic assumptions about something as simple as size can be extremely misleading. In this case, a bone that actually belongs to a group of very primitive catarrhines that originated before the apes and the Old World monkeys diverged from each other was incorrectly assigned to a genus that was not only an ape, but also a probable relative of the African apes and humans. This type of misidentification did, indeed, lead to an incorrect understanding of how our ancestors negotiated their forests. So we paleoanthropologists are really thrilled when we find skulls and postcranial (bones below the neck) remains together.

Unlike most fossil primates, *Ekembo* is known from almost every bone in its body, though not from any one complete skeleton. We know that *Ekembo* had arms and legs of roughly equal length and a spine with a long lower-back region that was carried parallel to the ground on which it stood. This is basically the same body plan as a living monkey, and in the trees at a certain distance, you would take an *Ekembo* for a monkey. Remember plate 2, which compares monkey and ape skeletons? For the most part, *Ekembo* falls among the monkeys; however, upon closer examination, you would notice a major difference from most living Old World monkeys: the absence of a tail.

For reasons that we do not fully understand, apes lack a tail. Apes also lack wings, but you would not define them from that perspective, so it is more scientifically meaningful to say that they have a tailbone (coccyx) rather than to say that they lack a tail. The coccyx is the end of the vertebral column in all apes, including humans; it curves inward, toward the front of the body. It probably helps to support the organs of the pelvic cavity (lower part of the gut or gastrointestinal tract, bladder, uterus in females, prostate in males, etc.). There are also ligaments that attach to the coccyx that serve as attachment sites for various muscles. So the coccyx is not just a vestige of a tail, it has a function; however, in most other primates, the coccyx does not exist, and other structures support those organs. Instead, these primates have caudal vertebrae, or tail bones, which curve in the opposite direction, toward the back, thus creating an external tail. *Ekembo* had a coccyx, and not a tail, and this is one of the most important traits that tell us that *Ekembo* is in fact an ape.[1] As I mentioned before, when you look at the early members of the Hominoidea, you expect that many of the distinguishing characteristics of living hominoids would not yet have evolved. This is the case with *Ekembo*, but the absence of a tail is a sure sign that it is an ape. Monkeys, all of which retain tails, diverged before the tailless common ancestor of *Ekembo* and living apes evolved.

Some researchers, and I count myself among them, think that the loss of the tail in *Ekembo* is related to its powerful grasping hands and feet. Based on the evidence of the bones from KPS, I concluded that *Ekembo* had more strongly developed grasping abilities than living monkeys. Tails serve an important function in agile, tree-dwelling animals: they allow them to easily shift the center of gravity and adjust their position in the trees. Agile arboreal primates, such as many monkeys and lemurs, have large tails that they move to shift their center of gravity and control exactly where they are going to land. If you do not have a tail, you have to manage movements in the trees with your hands and feet, which is a challenge. The absence of a tail may even have stimulated the development of those parts of the brain responsible for controlling fine movements of the extremities, especially the hands. One of my graduate students, Amber MacKenzie, has discovered a relationship between the development of the muscle of the hand and the length of the tail, suggesting that primates with shorter tails have more strongly developed grasping hands. This is an interesting idea because the area of the brain in the motor cortex that controls fine motor coordination of the hands is immediately adjacent to one of the central language areas of the brain. It is a bit of a stretch, but it is conceivable that the loss of the tail led to or was made possible by the development of greater degrees of control of the hands, which led to or required the development of that part of the brain that controls motor movement of the hands. Enlargement or reorganization of the hand region of the motor cortex may have facilitated or in a sense spilled over into the adjacent regions controlling such areas as the muscles of the face, throat, and tongue. So the loss of the tail more than 18 million years ago in *Ekembo* may have been, very indirectly, a necessary condition that would lead millions of years later to the evolution of an enhanced capacity to communicate by facial expressions and eventually to the evolution of spoken language. But as I said, this is very speculative.

Other than the presence of a coccyx and more powerfully grasping hands and feet than living monkeys, *Ekembo* lacked many of the characteristics of living apes. It did not have a broad chest with shoulder blades positioned on the back, and there are no indications that it could swing by its arms beneath the branches of trees. Its elbow is halfway between that of a monkey and an ape, but still more like a monkey's, and the same goes for its wrist, hands, and feet. For example, *Ekembo* has a large olecranon process. If you bend your arm at the elbow, bringing your hand toward your shoulder, you can feel your olecranon process, the pointy end of a bone on the outside of your elbow. If you now extend you elbow so that your arm is straight, it disappears. This is because your olecranon process, as well as those of all apes, is very short and fits into a pit behind your elbow (the aptly named olecranon fossa). The vast majority of quadrupeds cannot fully extend the elbow because the olecranon process protrudes well beyond the back of the elbow and does not fit into a pit. *Ekembo* was not able to extend its elbow fully (see plate 3).

While protrusion of the process may sound like a handicap for other animals, it actually serves them well. The olecranon process is the site of attachment of a group of muscles called the triceps humeri (known more commonly as the triceps). If you have a long olecranon process, your triceps has more leverage in elbow extension, and you will be a more powerful runner, though with less range of motion. All fast-moving mammals have long olecranons. So why reduce the length of the olecranon process? In hominoids the reduction leads to our ability to fully extend our arms at the elbow, which is critical for hanging (try hanging from a bar with your arms bent), and although it reduces our power, it increases the speed of extension at the elbow. It is without a doubt the reduction of the olecranon process along with changes in the anatomy of our shoulders and our amazing hand-eye coordination that allows humans to throw with great speed and accuracy, whether it's a spear or a 100-mile-an-hour fast ball in the strike zone. *Ekembo* retains the anatomy of an animal with an elbow adapted for powerful but limited range extension, which in the case of *Ekembo*, as in many arboreal animals, makes them excellent climbers—able to hoist themselves almost effortlessly into the upper branches of the canopy. While more modern apes have lost this feature of the elbow, they evolved new ways to retain their ability to move about in the treetops.

The wrist of *Ekembo* differs in a fundamental way from living apes and is more like that of other mammals. In most primates other than apes (and in fact in most other animals with limbs), both bones of the forearm attach directly to the wrist. The radius is on the side of the thumb (when your palm is facing outward) and is the major connection between the forearm and the wrist. It attaches directly to two bones of the wrist in all primates. I have known several people, including my son, who broke their "wrist" (my son did it twice), but in fact, it is the lower part of the forearm, usually the radius, that is broken. When a wristbone really breaks, it is usually the scaphoid, the larger of the two bones that attach to the radius. Whether it is the radius or the scaphoid, it is almost always the bones on that side (the thumb side) that break because they are the only solid attachments between the forearm and hand in humans and apes. In all primates except apes, the other bone of the forearm, the ulna, includes a projection called the ulnar styloid process that attaches to the opposite (pinky) side of the wrist. This helps to stabilize the wrist and makes a firm joint that does not deviate much from side to side. It functions like the hinge of a door, allowing for opening and closing but little twisting or rotation. In apes the ulnar styloid process is greatly reduced (like the olecranon), and it does not connect with the wristbones. So the ape wrist is much more mobile and less stable. It is, in other words, much more capable of twisting and rotation. So, why is this a good thing? It is not that it is good or bad per se, but that it works better for what apes do. Apes need to have more flexibility in their wrists in order to adjust the position of their hands as they grasp branches and negotiate the very complex, upside-down, three-dimensional environment of the trees.

By "upside down" I mean that apes brachiate—they hang suspended by their hands underneath the branches and swing their weight forward to catch the next branch. Apes move underneath the leafy canopy rather than on top of the branches, as do monkeys. So the apes' bodies pull on their hands and wrists, whereas the weight of the monkeys' bodies presses down on their hands. The absence of an ulnar styloid contact in the wrist probably helps to avoid injury, which would damage or break the relatively delicate ulnar styloid if there were excessive twisting of the wrist. Without an ulnar styloid, apes can twist their wrists and spin their bodies as they reach to grab the next branch. These are just two details in which *Ekembo* more closely resembles the common ancestor of apes and monkeys than apes themselves, even though it had already crossed that phylogenetic threshold onto the ape side.

Although the end of the vertebral column in *Ekembo* is modern in that it had a coccyx, above it *Ekembo* was monkey-like. As I mentioned, the lumbar region of *Ekembo* was long and horizontally oriented (see plate 3). The lumbar area is the part of the vertebral column along with the cervical or neck vertebrae that gives humans the biggest problems. In the neck the problem is that it is very mobile, in order to position the head where it needs to be to see things, and it supports a heavy object (our heads) on a fairly delicate structure. So we get neck pains and sometimes even ruptured disks. But by far the portion of the vertebral column that gives us the most problems is the lumbar vertebrae, or lower back.

I am often asked that if evolution leads to selection for the fittest, why do so many people have back problems? (One could also ask the same about special creation. Why did the Creator not take this into account when He was creating humans?) The answer, from an evolutionary perspective, is that all morphology in an organism depends on whatever the ancestors of that organism possessed. Fitness is relative, not a drive to perfection. Changes occur via the processes of natural selection, mutation, and random or chance recombinations, shufflings, and mixing of gene pools. Evolution is not a perfecting process, but one that leads to organisms that work in their environments as long as the environments do not change to the point where their anatomy no longer allows them to survive and reproduce. So what about our lower backs and *Ekembo*?

Ekembo has a lumbar region that is essentially the same as that of a monkey. There are seven lumbar vertebrae, and they are arranged in a nearly horizontal array. The lumbar region was flexible both due to its length and the anatomy of each of the vertebrae, which permitted movements between them, so that they had limber backs.

This is common in most mammals that need to adjust the positions of their hindquarters relative to their forelimbs or that need flexibility to bend and extend their backs during very fast running (check out https://www.youtube.com/watch?v=lGhTCPzMHvY—"Cheetah Run"— on YouTube to see how much a cheetah's back bends back and forth as it runs). Most mammals do this as well, though not to the extent seen in cheetahs. Apes have a reduced number of lower back vertebrae, as I mentioned in the introduction.

Ekembo was not adapted for suspension, though it was an excellent climber. Above the neck, *Ekembo* is also not immediately identifiable as an ape. The teeth are similar to those of *Aegyptopithecus*, which we have established are primitive for the catarrhines (Old World monkeys having evolved more specialized teeth), but there are subtle differences that convince most researchers that *Ekembo* is a hominoid. Without going into too much detailed terminology, the back teeth, or molars, of *Ekembo* look more like those of modern apes in being narrower side to side than they are front to back. They also have less pronounced ridges on the sides of the molars (figure 1.2).

The incisors also more closely resemble those of apes compared with those of most monkeys and seem to have been better suited for use in preparing foods (tearing off tough husks or peels, biting manageable pieces off for chewing, etc.) than in *Aegyptopithecus*. Like *Aegyptopithecus* and the vast majority of anthropoids, *Ekembo* was strongly dimorphic in canine size and shape, with the males having larger and taller canines than the females, even after differences in body mass are taken into account. While this suggests that *Ekembo* lived in social groups (again, like almost all anthropoids) and that there was a certain level of competition between males, we don't know the precise nature of the social organization of *Ekembo* (see plate 5 and figure 1.5).

The face of *Ekembo* is also primitive and lacks obvious signs that it is from an ape. It most closely resembles the faces of the lesser apes, gibbons, and siamangs, but all of these faces look very similar to those of short-faced Old World monkeys (e.g., *Colobus*). As you can see in figure 1.6, in cross-section the face of *Ekembo* resembles that of most other primates, including gibbons and siamangs. In

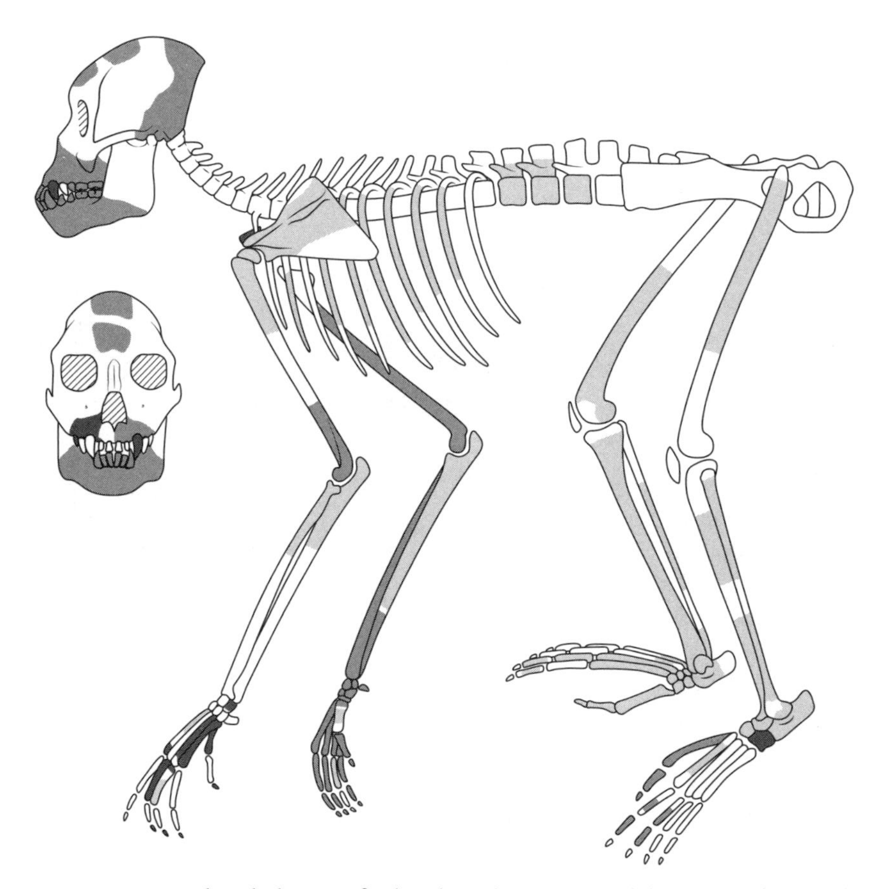

FIGURE 1.5. The skeleton of *Ekembo*. The arms and legs are of roughly equal length; there is a long, horizontally oriented back; and a tail is absent. (Modified from Walker and Teaford 1989.)

these animals there is a big gap between the part of the palate that houses the front teeth (the premaxilla) and the maxilla, the rest of the upper jaw. In great apes, the premaxilla is expanded, but in a different way in the line leading to orangutans than in that leading to African apes and humans. In orangutans the premaxilla, with its incisors, is large and sticks out more horizontally than in African apes. As you move from the incisors toward the nose, the premaxilla merges with the roof of the mouth, leading to a smooth transition from the base of the skeleton of the nose into the nasal cavity.

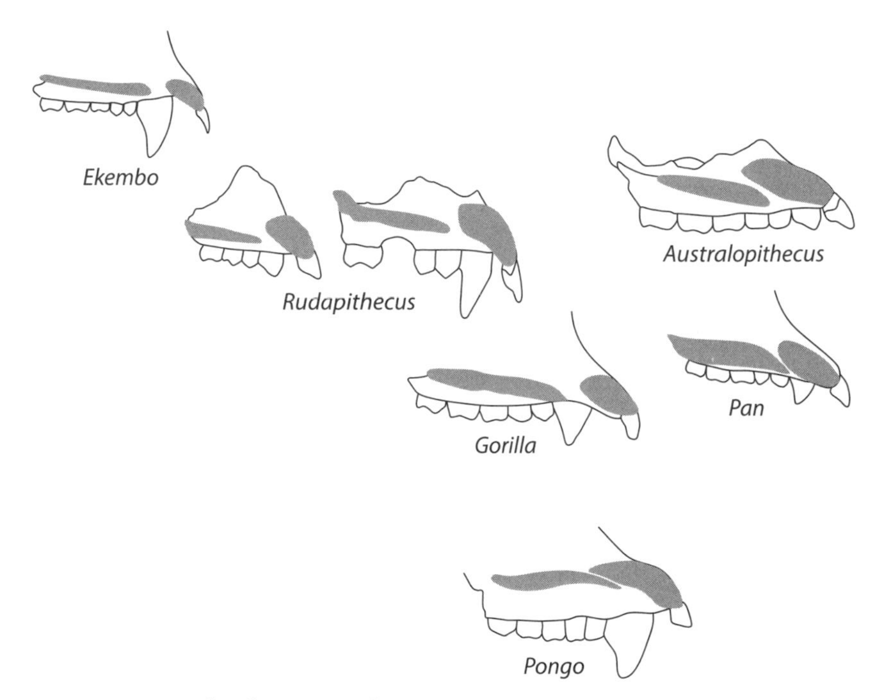

FIGURE 1.6. Palatal cross-sections.

This morphology contributes in part to the unique concave look to the orang in side view, which we will discuss in more detail in the chapter on pongine evolution. In African apes and humans and in a number of fossil apes, the premaxilla is enlarged, as well, compared with that in *Ekembo* and lesser apes, but it is less projecting and more vertical than in orangs, so that it does not overlap as much with the premaxilla. Instead, there is an overlap, but it is offset, resulting in what we call a stepped subnasal fossa (figure 1.6). *Ekembo* has a premaxilla that most closely resembles that of hylobatids and other non-hominid primates. Like *Aegyptopithecus*, *Ekembo* had eye sockets of a size that is consistent with a primate that was active during the day.

The brain of *Ekembo*, however, is unexpectedly large for a catarrhine of this geologic age. The brain of *Aegyptopithecus* is about the size that one would expect of a prosimian of the same body mass, that is, very small for a catarrhine but larger than any contemporaneous fossil primate. In contrast, the brain of *Ekembo* is equivalent in size, once body mass is taken into consideration, with that of gibbons and siamangs, and also baboons. Baboons have the largest brains, when scaled relative to body mass, of any Old World monkey. As a primitive ape, *Ekembo* has a brain that is relatively large, unlike *Aegyptopithecus*, which is a catarrhine but has a prosimian-sized brain. *Ekembo* was catching up in terms of brain size, having reached the small-hominoid threshold, but both *Ekembo* and *Aegyptopithecus* were brainy for their times.

I think this shows that even 20 million years ago, near the origin of the hominoids, there was already selection for larger brains. How do we know how large *Ekembo*'s brain is relative to body mass? Because we have a skull of *Ekembo* with a braincase well enough preserved to estimate brain size and many limb bones that allow us to estimate body mass. It is true that we do not have the skull and limb bones of a single *Ekembo* fossil preserved well enough to estimate brain and body size in the same individual, so there is a range of error in our estimates. What I am reporting here is our best educated guess. I think it will turn out to be very close, once we find a skeleton with a braincase.

Researchers have recognized three species of *Proconsul* and two species of *Ekembo* from sites in Kenya and Uganda. There is evidence that *Proconsul* or its descendants may have persisted much later in time, but as a relatively small part of the East African fauna. There are a number of specimens from sites in Kenya that date to about 12 million years ago, and another specimen, currently attributed to *Samburupithecus*, which is about 10 million years old and may represent the last surviving member of the *Proconsul* lineage. I will return to these fossils in chapter 9, when I take up the story of other apes from 10 to 12 million years ago in Africa that some researchers have linked to the origins of African apes and humans.

There are several other taxa in the early Miocene of East Africa that seem a bit more modern, that is, possibly more closely related to more advanced fossil apes and to living apes than *Ekembo*. Among these the most widely discussed are *Morotopithecus* and *Afropithecus*. Many researchers have concluded that *Afropithecus* is most likely to be closely related to later hominoids, and so I will discuss it in chapter 2. *Morotopithecus*, however, is the subject of greater uncertainty.

Morotopithecus, which is known from Uganda, is based on a large, well-preserved upper jaw that was originally attributed to *Proconsul major*, mainly because of its size. However, most researchers now consider it to be a different genus. There are a few features of the palate and teeth that distinguish *Morotopithecus* from *Proconsul*, but the main distinguishing features are in the limbs. We have two lumbar vertebrae of *Morotopithecus*, and the more complete one at least appears more modern than those attributed to *Ekembo* and more modern, in fact, than those in most apes from the middle Miocene (see chapter 4). This is curious because the jaws and teeth of *Morotopithecus* are more primitive than those of most middle Miocene apes and also because *Morotopithecus* is thought by most researchers to be about 20 million years old, roughly 5 million years older than the oldest known apes from the middle Miocene. Thus, it has a modern spine, primitive teeth, and old bones.

The vertebrae, shoulder blade, and hip joint of *Morotopithecus* suggest that it was capable of moving around in the trees with postures seen only in living apes. Apes tend to emphasize positions in which their vertebral columns (backbones) are oriented more vertically, both when they climb trees and also when they move on the ground. We refer to this as orthogrady, or orthograde posture. Humans are fully orthograde (our vertebral columns are completely vertical), but orthogrady alone is not enough to allow an animal to habitually walk upright, that is, to be a biped. Orthogrady in apes is related to their ability to hang and swing below branches (suspension), and that is what we may be seeing in *Morotopithecus*: a greater degree of apelike posture and locomotion than in *Ekembo*. Nonhuman orthograde and pronograde animals are all quadrupeds (they walk on all fours), not bipeds, like us, who walk on only two limbs.

One explanation for orthogrady in *Morotopithecus* is that the features that allowed it to move more like existing apes evolved independently from those that characterize late Miocene and today's apes. While it may sound like a stretch, orthogrady, and even brachiation, may have developed independently a number of times during primate evolution. Independent evolution of similar characteristics is so common that it is recognized as an evolutionary process. When this happens among closely related species that already share many anatomical attributes, as in the case of hominoids, we call it parallel evolution. When it occurs between more distantly related species with less in common anatomically, such as the suspensory habits of spider monkeys from the New World and gibbons, it is called convergent evolution. Large tree-dwelling apes are better off hanging from the trees rather than trying to stay balanced on the tops of branches. It is too hard for them to grip small branches tightly enough to avoid rotating off of them and falling to the ground. While there are a number of solutions to this dilemma, apes have opted to move below branches and rely on very powerful arms and hands to hang. *Morotopithecus* may have had an "incentive" to develop more suspensory postures independently. Being relatively large, *Morotopithecus* may have evolved a more vertical posture to allow it to move more comfortably in the trees, like chimps, for example, which are roughly the same size.

The face, jaws, and teeth of *Morotopithecus* are most like those of *Ekembo,* which walked with its vertebral column in a more horizontal position (pronogrady), like most mammals. While it is possible that *Morotopithecus* is more closely related to living apes than to *Ekembo* or the middle Miocene apes that I will discuss in chapter 2, I think the more likely explanation is that orthogrady and some degree of suspension evolved separately in this early, large, primitive ape.

Of course, one other explanation for the apparent ability of *Morotopithecus* to move more like an ape in the trees is that the specimens currently attributed to *Morotopithecus* actually comprise two different species. The skull fragments and teeth attributed to *Morotopithecus* may not belong to the limb bones, since they were not found together. But even if they do represent two different animals, we still are left with the problem of more modern-looking ape limb bones at such an early date.

There is one final little piece of uncertainty that adds to this mystery. The fossils of *Morotopithecus* were recovered over a very long time period, roughly from the 1960s to the 1990s. We are not completely sure where some of the fossils come from, and it is possible that they may be younger than thought by most researchers. Some researchers have suggested that they really date to 15 million years ago, which would make them the same age as the middle Miocene apes described in chapter 2. Even so, the limbs are still more modern looking than those of other middle Miocene apes.

I have a feeling that the *Morotopithecus* specimens do in fact belong to a single species and that they are probably 20 million years old, as the researchers working directly on this material have concluded; however, the other possibilities can't yet be ruled out. My best guess is that *Morotopithecus* was natural selection's early experiment in modern apelike posture and locomotion (which scientists shorten to "positional behavior"). These attributes and their associated behaviors evolved once in *Morotopithecus* and again, may be a few more times, in other lineages of apes, including our own. As I mentioned in the introduction, it is clear, given the multitude of similarities in the skeletons and apes and humans, that we passed through an orthograde and suspensory phase in our ancestry. The story of *Morotopithecus* and the parallel evolution of orthograde positional behaviors is one of those aspects of the fossil record of ape evolution that keeps things very interesting.

A review of the early record of hominoids from East Africa (they are only known from East Africa) would not be complete without mentioning the other early catarrhines that shared the landscape with *Ekembo*. As with *Ekembo*, just where they fit in the ape family tree is hotly debated. Along with the catarrhines from Rusinga I mentioned earlier (*Dendropithecus*, *Limnopithecus*, and *Nayanzapithecus*), there were a number of other apes on the scene (figure 1.7).

Micropithecus, as the name implies, is a very small early catarrhine, maybe three kilograms (not quite 7 lb.) or so in size, known mostly from isolated teeth and a well-preserved palate. It is from Kenya and Uganda and is about 19 million years old. At one time or another it was linked phylogenetically with gibbons (as were both *Dendropithecus* and *Limnopithecus*), but today researchers generally consider it to have an uncertain status (believe it or not, scientists have an official term for this, *incertae sedis*).

Turkanapithecus, which is known from the same site as *Afropithecus* (chapter 2) and not associated with any of the other primates discussed here, has an unusual morphology—a prominent snout and

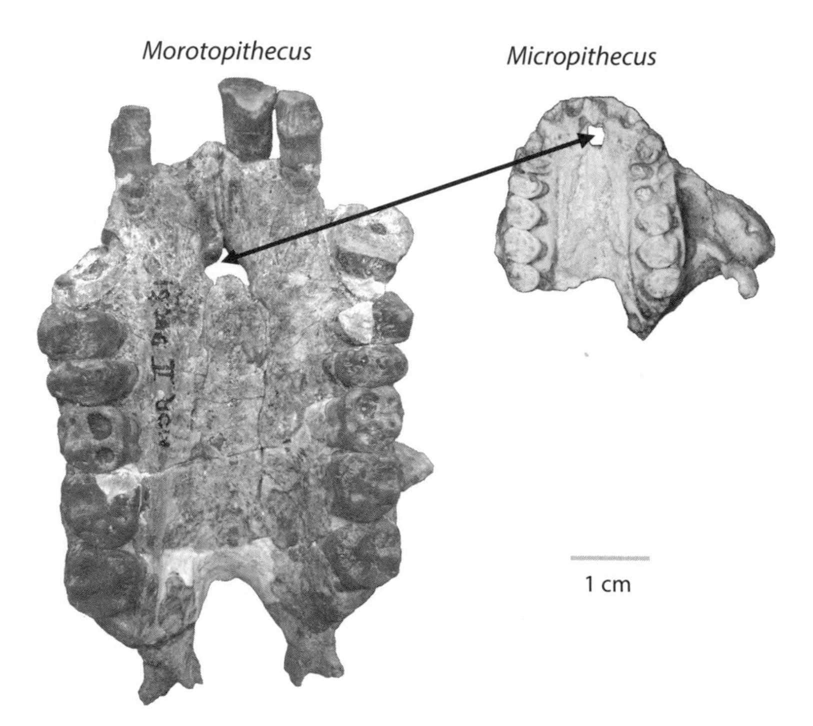

FIGURE 1.7. A comparison of the largest (*Morotopithecus*) and smallest (*Micropithecus*) of the early Miocene catarrhines. Note that despite the huge difference in size, both upper jaws have the same basic configuration, especially a large gap (*arrow*) between the front (premaxilla) and back (maxilla) parts of the palate. (Left image courtesy of Laura M. MacLatchy. Right image by author.)

teeth with complicated patterns of crests and basins. *Rangwapithecus* is known from a site in Kenya that is dated to about 19 million years ago. It is about the same size as the smaller species of *Proconsul* but has very distinctive tooth anatomy. Its teeth have strongly developed crests and tall, pointy cusps. Some have described the canines as scimitar-shaped, that is, tall and flat with a very sharp back edge. No one knows what *Rangwapithecus* did with these impressive canines.

Simiolus, also found with *Afropithecus* and *Turkanapithecus* at Kalodirr, is known as well in younger deposits (15 million years old) elsewhere in Kenya. *Kalepithecus* is also from Kenya but is older, about 19 million years old. Both of these apes were similar in size to *Limnopithecus* and are distinguished by the relatively detailed attributes of their teeth. We just do not know how all of these generally small catarrhines are related to later apes. It could even be that the real ancestor of modern apes is hiding among them, unrecognized because of its small size, and we have thus far unearthed very few limb bones. However, I think it is more likely that this impressive array of catarrhines represents an early adaptive radiation of more advanced primates into the available niches of East Africa during the early Miocene.

Darwin's finches and tortoises in the Galápagos Islands are classic examples of adaptive radiation. Adaptive radiation happens when an organism disperses into a new setting with few predators or competitors and rapidly evolves into a diversity of species, each occupying a different available niche in the new environment. Darwin's finches, for example, found themselves isolated on the islands of the Galápagos chain with little competition and then evolved into a number of different species, each with its own manner of obtaining food. George Gaylord Simpson, perhaps the most prominent American vertebrate paleontologist of the twentieth century, described the adaptive radiation in South America of many different kinds of mammals, including New World monkeys, with the phrase "a splendid isolation."

Populations of the early not-quite-apes from Africa would have been small and isolated, giving natural selection a chance to act. Some of the small catarrhines for which we have bones other than skulls and teeth, such as *Dendropithecus*, appear even more primitive than *Ekembo*, but others, such as *Turkanapithecus*, more closely resemble *Ekembo* and *Proconsul*. Until we find more and better-preserved fossils of these small catarrhines, we can only speculate on their place in the broader patterns of ape evolution. In the end the big picture is that in the early Miocene, many natural experiments in being a catarrhine occurred in East Africa. These resulted in a wondrous diversity of animals resembling the diversity of living monkeys. It was the first "golden age" of the apes, or near-apes. Another would occur in the late Miocene, as we will see later.

CHAPTER 2

OUT OF AFRICA: *AFROPITHECUS* AND FRIENDS

About 17 to 17.5 million years ago, in the northern part of Kenya, far from the sites around Lake Victoria from which *Ekembo* and most of the early catarrhines are found, a new type of ape appeared. *Afropithecus* is found at several sites, but the best collection, both in terms of numbers of specimens and their preservation, is from Kalodirr. A few other small catarrhines are also known from Kalodirr, in particular *Turkanapithecus*, but *Afropithecus* seems to have the best chance of being related to later apes.

Although the area is a desert today, *Afropithecus* like *Ekembo* lived in a forested environment. Many of the same animals found at Rusinga are also known from Kalodirr. We know that the fauna was dominated by forest-dwelling animals, and this is confirmed by the large number of fossil plant fragments (leaves and pieces of wood) known from the site. Several species of tragulid, the forest-dwelling mouse deer, are known, but there are also primitive giraffes, elephants, and those odd deinotheres.

When *Afropithecus* was first described in the mid-1980s, researchers were struck by a number of unusual features. The type specimen of *Afropithecus* includes most of a face, and it is by far the best specimen.[1] It has defined the morphology of the taxon in the minds of researchers since the 1980s. The face is very flat from above the eyes to the incisors. By "flat" I mean that in profile there is a straight line connecting the forehead and lower parts of the face, but this line is strongly slanted upward (see plate 6). In some ways this resembles one of the best-known skulls of *Aegyptopithecus*, although other *Aegyptopithecus* specimens are less extreme.

This is a very unusual facial configuration for a large catarrhine. In most anthropoid skulls, the eyes face forward. As you can see in plate 6, in order for the eyes to be positioned in a realistic way in *Afropithecus*, the face has to be strongly tilted, resulting in a tooth row that is very strongly inclined upward from front to back. This is possible, and in fact we see this in baboons, but baboons have extremely elongated faces, and other Old World monkeys lack this configuration. No other ape looks like this.

In addition, the *Afropithecus* face has a very impressive snout, again resembling *Aegyptopithecus*. Most catarrhines, even those with long faces like baboons, have small snouts; that is, their olfactory organs are reduced. Baboons appear to have large snouts, like dogs (in fact they are sometimes called dog-faced monkeys), but this is really an illusion caused by their massive canines. The roots of their upper canine teeth are so large that they push the middle part of the face forward. The skeleton of the nose is, in fact, far behind the front of the jaw. In animals with true snouts that house elaborate olfactory organs, the nose projects forward to the level of the front of the face. *Afropithecus* appears to have an intermediate condition, with a large snout for an anthropoid but a smaller one than *Aegyptopithecus* or prosimians. The *Afropithecus* face also has a huge space between the orbits (eye sockets), and these orbits are very small. The space behind the eyes, the postorbital constriction, is extreme, pinching off the face from the braincase. So, overall, *Afropithecus* is a very unusual-looking ape.

It has been suggested that this unique morphology may have been characteristic of the last common ancestor of Old World monkeys and apes, because it is also found in the earliest known Old World monkey, *Victoriapithecus*. It is quite unlike *Ekembo* in many ways. However, there is another possible explanation for the unusual look of *Afropithecus*, as we will see a bit further along.

Dentally, *Afropithecus* looks more "normal." The molars of *Afropithecus* are distinct from those of *Ekembo* in having larger, somewhat more rounded cusps, but they are not dramatically different. They may have a somewhat thicker layer of enamel, the hard outer covering of the teeth, but this is not completely clear. The premolars are more distinctive, being very broad from the check to the tongue side. The canines are the most distinctive and have been described as tusklike; that is, they are very thick and relatively short. The incisors are also very thick and project forward strongly (procumbent) in the one specimen that preserves the face (but see below). Overall, the teeth of *Afropithecus* suggest an animal that used its back teeth for crushing and grinding hard or tough foods and its stout front teeth for removing tough husks or other outer coverings from fruits or other foods (nuts, roots, etc.). This way of feeding and the choice of foods is different from those of *Ekembo*, but we find examples in living primates, particularly in some South American monkeys that we call sclerocarp feeders—that is, they feed on fruits covered by tough protective skins.

Other aspects of the skull of *Afropithecus* are consistent with this feeding strategy. The areas that are generally reinforced in animals that have very powerful bite forces are also strongly developed in *Afropithecus* compared with *Ekembo*. *Afropithecus* has prominent cheek bones, to which large masseter muscles attached (the muscles that bulge from your cheeks when you chew), and strongly developed temporal lines, for the attachment of a powerful set of temporalis muscles (the muscles that bulge behind your eyes when you chew forcefully). The mandible is also reinforced along its length to withstand very high bite forces. All in all, the feeding adaptations of *Afropithecus* are clearly different from any other primate from the early Miocene.

But what about the highly unusual morphology of the face of *Afropithecus*? Its face appears to be highly specialized, and it is legitimate to wonder if it is likely to be the ancestral condition of descendants with much more typical-looking ape faces. That is, although the face of *Afropithecus* has been likened to those of early primates such as *Aegyptopithecus* and the ancient catarrhine cousins of the hominioids, such as the early Old World monkey *Victoriapithecus*, it does appear to be specialized for a specific kind of diet. So it is a bit weird that the much more typical or generalized-looking ape faces of *Ekembo* and other early Miocene apes could have evolved from an *Afropithecus*-like ancestor.

I had a chance to reexamine the original fossils during the summer of 2011. I had seen the originals shortly after they were described, in the late 1980s, and I had no issues with the assessment of the describers, Alan Walker and Meave Leakey, two luminaries in the field. I did think that the face was odd, especially the orientation of the eye sockets relative to the tooth row, but I put that down to the fact that it is a fossil ape, and so why should it necessarily follow all the rules of living ape skulls. Almost twenty-five years later, I noticed something that had escaped my attention earlier—that most of the areas that are the most unusual in the face of *Afropithecus* are actually not preserved or they are damaged. A number of regions of the face of the type of *Afropithecus* are represented only by matrix, the sediment that fills cavities in bones and hardens, in this case, into a material with the consistency of concrete. The premaxilla and most of the area around the nose is actually matrix and not fossilized bone. Even some of the teeth have been replaced completely by matrix, which is something I have not seen elsewhere. In my opinion the face has been pulled out during the process of fossilization, and it has also been compressed, resulting in the unusually large space between the orbits, the smallness of the orbits, and probably the extreme postorbital constriction. Undistorted, this specimen of *Afropithecus* probably looked much more like an *Ekembo* than we ever thought.

There are several important lessons here. First, look at the original specimens and not just casts. Casts are extremely useful, and I have spent a great deal of time molding hundreds of fossil ape specimens, but they cannot replace examination of the originals. It is impossible to see this deformation on a cast, no matter how high the quality. Second, keep an open mind. Anything is possible. When I first examined the specimen in 1987, I was looking for what had already been described. I was not critical enough.

If the idea about distortion is correct, it is an unusual pattern of deformation, but not unheard of. Most important, though, it alters this part of the story to a significant degree. *Afropithecus* is not so much an entirely new kind of fossil ape poised to expand its range into Eurasia as it is an ape not too dissimilar from *Ekembo*, but one with modifications to the jaws and teeth that allowed it to exploit a broader range of resources and thus to expand into new environments.

What is definitely different in *Afropithecus* compared with *Ekembo* is its powerful chewing apparatus. The strong attachments for the chewing muscles (temporalis and masseter) are real, as are the robust incisors and canines. It is not clear if the incisors were really as procumbent in life as they appear on the fossil, so it may not have been a true sclerocarp feeder. However, there is no doubt that *Afropithecus* could access and process foods not available to *Ekembo*, perhaps giving it more dietary breadth and allowing it to expand its range over time.

RECOVERING LIFE HISTORY FROM TEETH

Another aspect of the biology of *Afropithecus* that seems to differ from *Ekembo* is the way it grew. In the last twenty years or so, a mini-revolution has taken place in our ability to deduce aspects of the biology of extinct organisms that no one would have believed possible. It has been known for some time that different primates grow at different rates and that their growth is associated with a number of important aspects of their biology, traits referred to as life history characteristics. For example, all great apes erupt their permanent molars at roughly 3, 6, and 9 years of age for the first, second, and third molars respectively. We humans, on the other hand, erupt our molars at 6, 12, and 18 years, about twice the time it takes great apes. Interestingly, australopithecines, the ancestors of our genus (*Homo*), seem to have close to same timing as apes, perhaps only slightly slower in some taxa. Even early *Homo*, such as *Homo erectus*, had a faster rate than modern humans, but it much closer to that of humans than of apes. It turns out that in primates, the age of eruption of the first molar correlates with a number of attributes, including brain size and several aspects of life history. Surprisingly, we can make a well-informed educated guess at the brain size category (lemur-like, monkey-like, apelike) from the age at which the first molar emerges. The correlations vary in strength, but in general, the age of eruption of the first molar is correlated with such life history traits as life span, age at menarchy (reproductive age in females), lifetime number of offspring, size of the newborns relative to the mother, and others. So if you can determine the age of first-molar emergence in a fossil taxon, which is possible but challenging, you have a strong basis for characterizing attributes of the biology and behavior of fossil species that can never be deduced directly from morphology.

So how can we determine how *Afropithecus* grew and how it might have differed from *Ekembo*? We still cannot grow extinct taxa in test tubes (although we are getting close for mammoths preserved in the Arctic tundra), so we need to use indirect methods. It turns out that teeth in mammals encapsulate the growth process, and this information can be teased out of them, although with much patience. The crowns of teeth consist largely of a cap of enamel overlying a base of dentine. Dentine is essentially the tooth version of bone, but enamel is chemically different, being the hardest substance the body produces. This is the reason that teeth are by far the most common of all vertebrate fossils.

Enamel preserves very well, and its internal structure is very often perfectly preserved. Enamel grows out from the junction between the enamel and the dentine (the enamo-dentine junction, or EDJ). The cells of the developing tooth that secrete enamel (the ameloblasts) move out in a sort of growth front comprising hundreds of structures called prisms toward the eventual surface of the tooth. The cells appear to secrete enamel in daily increments, represented by a slight thickening in the prism. Imagine squeezing toothpaste out of the tube. The toothpaste comes out in a sort of blob that tapers as you pull away the tube. Each day a new blob of enamel is secreted, and each of these blobs is separated by a constriction that appears under the microscope as a line, called a cross-striation. The ameloblasts seem to pause at regular intervals, for reasons that are not known, and this inactivity results in stronger brown lines, known as striae of retzius (figure 2.1).

Each species has a characteristic number of daily deposits of enamel between each brown line, which is known as its periodicity.

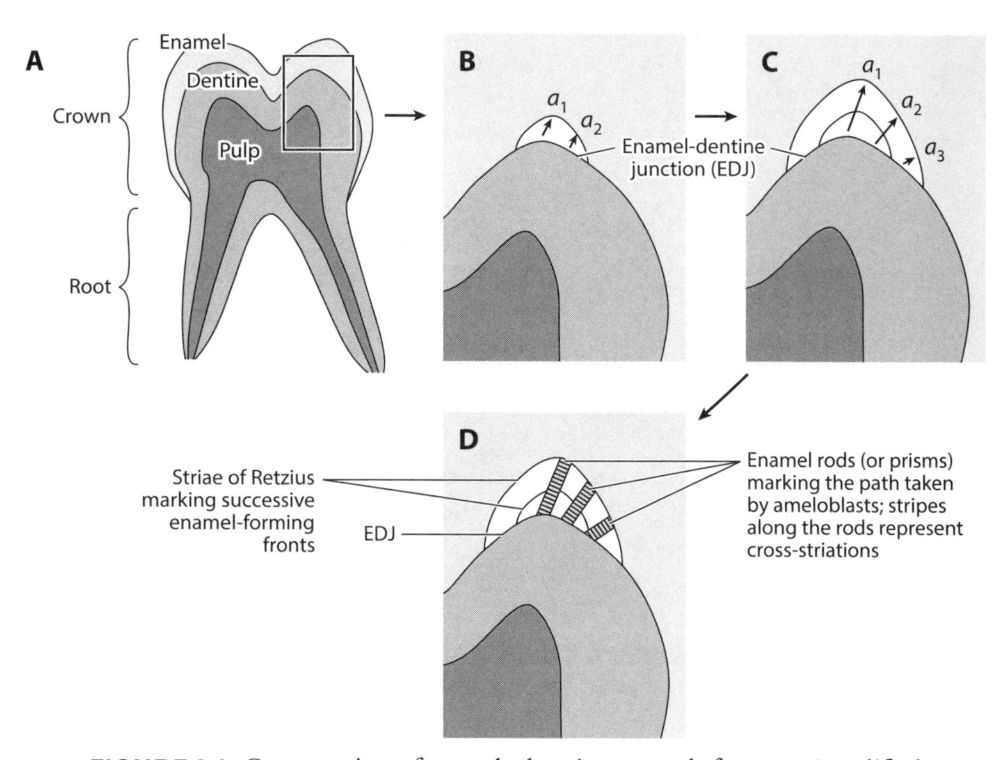

FIGURE 2.1. Cross-section of a tooth showing growth features. (Modified from Guatelli-Steinberg 2010.)

In great apes and humans, the periodicity on average is seven, which is taken to represent seven days. In monkeys and lemurs, for example, the periodicity is lower, indicating less time between pauses, which results in a more rapidly developing tooth.

In order to reveal the periodicity in the tooth of a particular species you need to have a very clean cross-section of the enamel cap, either a physical cross-section, like a histological section of tissue used to diagnose diseases, for example, or a virtual cross-section, obtained by an imaging device. By "clean" I mean a cross-section in which you can clearly see all the structures you need to count, which is more difficult to get than you might think. Once you have a usable cross-section of a tooth, whether a physical or a virtual slice, you need to count the number of cross striations between the brown lines in several spots, preferably, on the cross-section to get the periodicity. This is painstaking work that can consume many, many hours, and of course, it has to be repeated a few times to be confident in the accuracy of the counts. But what we learn from this effort is so important that it is well worth the time and effort. Once you have the periodicity, all you need to do is count the brown lines and multiply by the periodicity, and you have the amount of time it takes the crown to form. I say, "All you need to do . . . ," but the entire process is extremely labor-intensive, especially if you want to be able to compare one species to other. It's the perfect job for a graduate student.

To get the most important piece of information— that is, the age of eruption of the first molar—you need to know the crown formation time, but also the age at which the crown began to form and the amount of time that passed between the completion of crown formation and the actual eruption of the tooth. Crowns remain covered under the gums for some time after they are completed and emerge only following the growth of a certain amount root. Fortunately, we can estimate the amount of time since the root began to form by its length, that is, if it is not damaged. We also know that in all apes living today, as well as in humans, the first molar crowns actually begin to develop about one month prior to birth. So if you have a fossil specimen with the first molar just erupted (which you can tell by its position in the jaw and the amount of wear on the cusps), all you need to do is remove the tooth, calculate the crown and root formation times, and add a month, and you have the age at first-molar emergence and, of course, also the age at death of the individual. It is a process that would make any CSI lab proud.

So to summarize, the growth of the enamel cap in a primate tooth leaves a record, analogous to tree rings, that allows us to determine how much time it took for the enamel part of the crown to form. To get the most accurate counts, you really need to see all the striae, and of course you need to see the cross striations clearly. Until recently this was only possible by examining thin sections, that is, cutting the tooth, polishing the cut surface, mounting a slice on a slide, and then getting down to counting under a microscope.

This is a destructive technique. Even though the slice is only microns in thickness and the tooth can be repaired so that there is no sign of the cut, many curators of fossil primate teeth refuse to allow researchers to cut up their precious teeth. Of course, ideally you also need to find the jaw of an individual that died at the tender age of first-molar eruption, where there is no wear on the tooth. Since fossilized baby apes that died around the time of eruption of their first permanent molars are quite rare, not even the most liberal museum curator would allow a researcher to extract a tooth and slice it in half. So researchers have found a somewhat indirect but still quite reliable way around this problem.

As I mentioned earlier, it is possible to use a virtual cross-section to count cross-striations, but the only device I know of capable of producing images of the inside of a tooth at sufficient resolution to see cross striations clearly is a synchrotron. A synchrotron is not something you would find in a typical university lab, or for that matter, in typical university. It is a massive particle accelerator designed for experiments involving elemental particles, with numerous applications from fundamental research in physics to applications in industry. The X-rays produced by a synchrotron are hundreds of times brighter and more intense than conventional X-rays and can penetrate deep into highly mineralized, dense materials. These multimillion dollar installations, usually housed in huge underground facilities, were not originally designed for paleoanthropologists to peer inside a tooth, but fortunately there is occasionally room to have fun with these otherwise very serious machines.

In a few places around the world, paleontologists have had their specimens scanned, resulting in a tremendous increase in our ability to extract information from fossils. Insects embedded in amber, dinosaur embryos still in their eggs, skulls of the earliest-known primates, and teeth from fossil apes and early humans are among the important paleontological specimens that have been scanned in recent years. I have been fortunate to have been able to have fossils from my project in Hungary scanned at the European Synchrotron Radiation Facility (ESRF) in Grenoble, France.

In the case of *Afropithecus*, researchers were able to use a synchrotron to examine a baby's jaw with the first molar just coming out (but not quite yet erupted) and some isolated teeth that they were able to section. The teeth gave them the periodicity and crown formation times, and using estimates based on living species, the researchers were able to estimate root formation times, and thus the age at death of this little *Afropithecus*. So, what is the difference in dental maturation between *Afropithecus* and *Ekembo*? *Afropithecus* has longer crown-formation times and an older age at first-molar eruption than *Proconsul* on average. Whereas this age in *Ekembo* is similar to that in many monkeys, such as baboons, and also gibbons, in *Afropithecus* it appears to be intermediate between that of monkeys and great apes. From this we deduce that *Afropithecus* had a slower life history. And among other things, it probably means that *Afropithecus* had a somewhat larger relative brain size and possibly a longer period of infant dependency and learning. Along with their powerful jaws and ability to exploit a wider range of food resources than *Proconsul*, the additional trump cards of bigger brains and longer childhoods with more time to learn from their parents may have allowed *Afropithecus* to expand its range out of Africa, as we shall see in chapter 3.

LEAVING AFRICA: FOSSILS FROM THE ARABIAN PENINSULA

Some fossil evidence supports the scenario in which *Afropithecus*, with its dietary and life history innovations, may have in fact been the ancestor from which the first hominoids that lived outside of Africa evolved. There is a site in Saudi Arabia that is very close in age to Kalodirr. Here, a place called Ad Dabtiyah, jaw fragments and teeth were discovered from a new fossil ape (see figure 1.1) in 1978. *Heliopithecus* is the oldest ape found outside of Africa, or, I should say, geopolitical Africa. The Saudi Arabian peninsula is actually part of the geological continent of Africa, and at the time it was not a vast desert but rather a forested environment similar to those in East Africa in which apes are found. In fact, at this time, at the end of the early Miocene, the tropical zone was much wider, both northward and southward, and the climate was considerably wetter and warmer as far north as Germany. The climate was also one of the factors that allowed the apes from Africa to disperse north into Europe, as we shall see in chapter 3. So it is not really that surprising that we would find an ape in Saudi Arabia, and it is revealing that this ape happens to be one with powerful jaws and teeth adapted for crushing food and withstanding high chewing forces. It closely resembles *Afropithecus*, although sadly it is only known from a few teeth and a severely crushed upper jaw. In fact, some researchers include both apes in the same genus. Whatever taxonomists eventually conclude, *Heliopithecus* and *Afropithecus* are close relatives. *Heliopithecus* is in exactly the right place and time that we would expect for an ape that had expanded its range out of Africa at the end of the early Miocene, and its morphology strongly suggests that *Afropithecus*, or a close relative, was this ape. Also revealing is the fact that this is exactly the kind of ape that we will encounter in the next act of this evolutionary play, the origin of our evolutionary family, the Hominidae.

CHAPTER 3

OUT IN THE WORLD: EARLY APES SPREAD IN EUROPE

About 17 to 17.5 million years ago, an instant in geological time after *Afropithecus*, its possible descendants appeared on a new stage of hominoid evolution: Europe. This new kind of ape is called *Griphopithecus*, or the griphopiths (including a few species, as we will see below.) "Griphopiths" is not a technical term. It comes, obviously, from *Griphopithecus*, one of several apes mostly from the middle Miocene that all have thicker, more massive jaws than *Ekembo*. The teeth are also more modern, having reduced cingula (those ridges on the sides of the molars), low rounded cusps, and thicker enamel. However, the most surprising thing about these new apes, the griphopiths, is where they are found. Griphopiths first appear in the fossil record in Europe, and they are the first hominoids to be found out of Africa. So where did they come from and how did they get to Europe?

As I mentioned in chapter 2, I think the most likely candidate for ancestor of the griphopiths is *Afropithecus* or a related form. I use the caveat "a related form" because we can never be sure, despite frequent claims to the contrary, that a fossil taxon is the actual ancestor of another taxon. It could well be that *Afropithecus* is a side branch and the actual ancestor of the griphopiths is a relative of *Afropithecus* that has yet to be discovered. This is a bit of a phylogenetic trick to keep our options open. After all, when you consider how difficult it is to distinguish among closely related species today, it would not be too surprising if we have failed to properly identify all the species actually represented by the fossil record.

We cannot compare the DNA of apes that are millions of years old, yet sometimes this is the only way to distinguish among species today. Let's look at a group of closely related living apes. Not long ago, only one genus with three species of living gibbons was recognized, based mostly on size, which we can assess in the fossil record. These were the siamang (*Hylobates syndactylus*), the white-handed gibbon (*Hylobates lar*), and the crested gibbon (*Hylobates concolor*). Several subspecies of the latter two were also recognized, distinguished by "soft" characters such as fur color and song attributes (yes, gibbons sing), which we cannot find in the fossil record. Today we recognize four genera, each with a different number of chromosomes! There are at least 6 species in the "lar" group, 4 in the "concolor" group (now known as *Nomascus*), and a third genus (*Bunopithecus*), for a total of 11 gibbon species. Plus, we now recognize the siamang as a separate genus (*Symphalangus*). So based on genetics, as well as on more in-depth analysis of differences in morphology, fur color, singing, and other behavioral attributes, we have gone from 3 species of gibbon to 12! This without discovering a single "new" species in the wild.

Closer to home, our taxonomic concept of the African apes has also changed over the years. Traditionally, three species were recognized. These were *Gorilla gorilla* (the gorilla, obviously), *Pan troglodytes* (the common chimpanzee), and *Pan paniscus* (the bonobo, formerly the pygmy chimpanzee.) Some researchers even advocated placing all the African apes in a single genus, the genus *Pan*. Today, based on genetic, behavioral, and in-depth morphological analysis, there are still only two genera, but many researchers recognize multiple species. In addition to the bonobo, two species of chimpanzee are widely recognized: *Pan verus* is the western masked chimpanzee, and *Pan troglodytes*, which is subdivided into two subspecies, *Pan troglodytes troglodytes*, the central African black-faced chimpanzee, and *Pan troglodytes schweinfurthi*, the eastern long-haired chimpanzee, the latter made famous by Jane Goodall. Many primatologists split the genus *Gorilla* into three separate species: *Gorilla gorilla*, the western lowland gorilla; *Gorilla graueri*, the eastern lowland gorilla; and *Gorilla beringei*, the mountain gorilla, the "gorillas in the mist" made famous by Dian Fossey. A few primatologists have recently suggested that there may be a fourth species, *Gorilla diehli*, from the Cross River region between Cameroon and Nigeria. Once again, without discovering a new population in the wild, more in-depth analysis revealed the presence of more species and varieties than previously suspected.

In addition to this problem, we know that the fossil record, as wonderful as it is, is incomplete. We are still discovering new species of living animals today, including primates. Recent additions to our family include the bushy-bearded titi monkey, discovered in South America in 2010; and the blue-testicled vervet, discovered in Africa, and three new species of loris from Borneo, including a venomous one, all discovered in 2012. While is it not easy to be confident that we have identified all living primate species, given that some inhabit the relatively unexplored areas of the densest and most inaccessible forests, it is even more difficult to know how many fossil primates are missing from our known inventory. As I paraphrased Darwin in the introduction to this book, the fossil record is like a book most of which has been erased by time, and it is a virtual certainty that we have not and probably will never find fossils of more than a small fraction of the species that have ever lived. In many cases it might even be less than 5%. Even if we do know of half of all the extinct apes that lived in the Miocene, and I doubt we know that many, there remains a strong probability that we have not found the ones that are the direct ancestors of apes that evolved after them. A final thought about the incompleteness of the fossil record: we know more than fifty genera of extinct apes. If this is half of the actual number that existed in the past, that would make over 100 genera, compared with 4 genera of living hylobatids and 4 genera of hominids (great apes and humans.) With a minimum of 100 genera of fossil apes in the Miocene we are talking about the *real* planet of the apes.

Returning to the griphopiths, in paleontology it is not rare for revisions of species to conclude that two different species or even genera were mistakenly classified as a single taxon. This is actually what happened for one griphopith and in several other cases described in chapter 4. That being said, something closely resembling *Afropithecus* or *Heliopithecus* probably gave rise to the first griphopiths.

I speculate that the robust jaws and powerful chewing muscles of the "afropiths" (an informal term for *Afropithecus* and *Heliopithecus*), possibly along with changes in their life history, enabled them to exploit a wider range of resources than the proconsuls that lived at the same time. Proconsuls have a chewing mechanism broadly similar to that of a chimpanzee, one designed to process relatively soft foods. Fruits with very tough outer coverings or hard foods, such as nuts, may have represented a problem for the proconsuls. Underground resources, such as tubers, may have also been a challenge for *Proconsul* and *Ekembo*, because they tend to be fairly tough but especially because they are covered in grit, which would quickly wear down their teeth. *Proconsul* and *Ekembo* would not have been able to extract adequate amounts of nourishment from leaves, because their teeth lack the sharp ridges that allow folivores (leaf eaters) to finely chop leaves, slicing through the cellulose and releasing the more caloric and nutritious cell contents. In addition, folivores tend to be large compared with closely related frugivores, because folivores need to ingest large amounts of vegetation in order to get an adequate return, and so they need big guts. This is one of the reasons gorillas are larger than chimpanzees.

Afropiths also lack the adaptations for eating leaves, but their powerful jaws and large front teeth probably allowed them to more efficiently process tough and hard foods. Like most primates, the afropiths probably preferred sweet, juicy fruits, as do gorillas today, but when those preferred food items were not available, they were able to fall back on less desirable but still nutritious foods, again, much as gorillas do today with leaves. In fact, fallback food strategies have recently been recognized as an important adaptation in many primates, perhaps even the major driving force behind selection for specific dental adaptations. You do not need special adaptations to eat a banana, but you will not survive, or you will not be able to feed your offspring even if you do survive, if you cannot process more challenging resources when bananas are not around.

When fruits become scarce, gorillas fall back on highly fibrous foods, like plant stems and bamboo, for which we use the appetizing term "terrestrial herbaceous vegetation" (THV), although mountain gorillas rely more routinely on THV than other gorillas do. Chimps also use THV, in addition to bark, figs, and leaves, while bonobos depend more heavily on THV. Orangs generally eschew THV in favor of bark, leaves, and in some cases figs, while gibbons and siamangs focus on leaves and figs. However, it is the gorilla that seems to rely most on fallback foods, and this reliance has led to apparent adaptations of the dentition to process THV, such as teeth with strongly developed shearing crests and tall pointy cusps. In the case of the afropiths, their large front teeth and massive chewing apparatus may well indicate an equally strong need to have an efficient fallback feeding strategy, which is probably what allowed the afropiths to expand their ranges north, first into Saudi Arabia, and then into Europe and Anatolia (Asian Turkey).

THE FIRST GRIPHOPITHS

The first apes to find their way into Eurasia initially expanded their range northwest from East Africa and Saudi Arabia into Europe. Although the oldest ape fossil known is from Germany to the west, other sites for such fossils are located on the shores of an ancient inland sea called the Central Paratethys. The Central Paratethys was part of a massive seaway, the Tethys Sea, which stretched from the Atlantic Ocean to the Indian Ocean and separated Africa from Eurasia. It was the barrier that prevented the movement of land animals between these two land masses until the early Miocene, when it began to break up. As Africa rotated counterclockwise and collided with Europe, the Tethys Sea broke up to form the precursor of the Mediterranean and of an inland sea and lake system known as the Paratethys. In time, the Paratethys would itself reduce in size and break up. The western part of the Paratethys disappeared first, whereas the Central Paratethys lingered until about 6 million years ago. The Eastern Paratethys still exists. Today we know it as the Black and Caspian Seas.

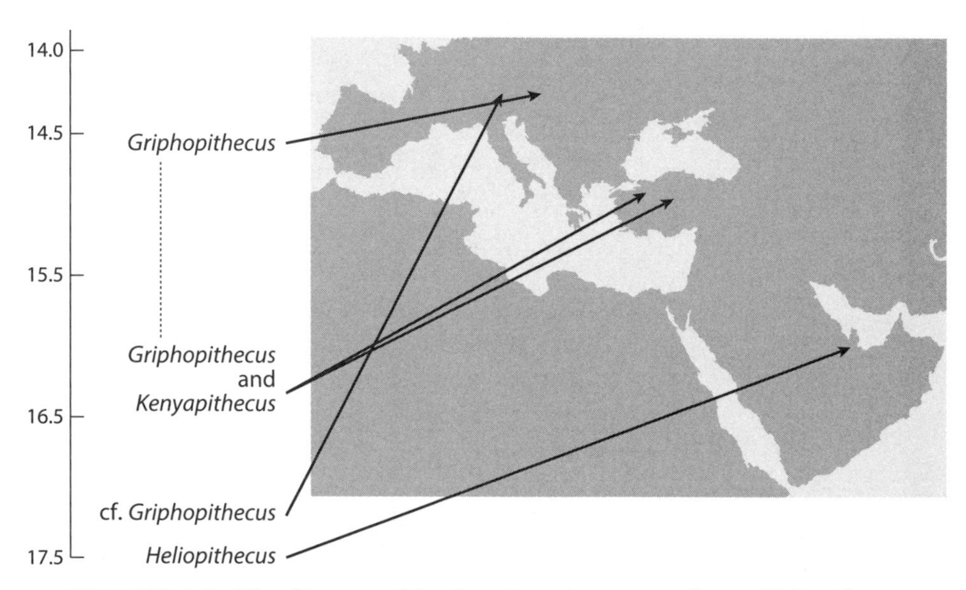

FIGURE 3.1. Timeline graphic showing time span from *Heliopithecus*, Saudi Arabia (ca. 17.5 Ma) through the middle Miocene (ca. 14 Ma), with a map showing where the fossils were recovered (Germany, cf. *Griphopithecus*; Turkey, *Griphopithecus* and *Kenyapithecus*; Slovakia, *Griphopithecus*).

The Central Paratethys is also known as Lake Pannon. Pannonia was a province of the Roman Empire that included Hungary and parts of Austria, Croatia, Serbia, Slovenia, Slovakia, and Bosnia and Herzegovina. The large geomorphological structure known as the Pannonian Basin is bounded to the north and east by the Carpathian mountain range (Slovakia, Romania), to the south by the Dinaric mountain range (Croatia), and to the west by the Alps (Slovenia, Austria) (figure 3.1).

The deposits are deepest however in Central Hungary, where sediments from the ancient lake reach several kilometers in thickness. The bottom of Lake Pannon is several kilometers under the current land surface, mainly because the entire region has been subsiding, or sinking, as the sediments accumulated. There is therefore no hope that we will ever see much of the ancient lake at the time that apes lived on its shores. Fortunately, at several places at the edge of the lake when it was at its largest extent (roughly the size of the modern

Caspian Sea today), fragments of shoreline have been preserved. When we find fossil apes in the Pannonian Basin, they are always found on these ancient shores, which suggests that there was something about this type of environment that pleased the apes, allowing them to thrive. It may well have represented a local ecological zone lush with plant life and its microclimate buffered to some extent by the effects of the lake. It is also a handy way of giving us a starting point for looking for new ape sites. In fact, with colleagues we have started to search for new sites on these preserved shorelines in Romania and Croatia.

By the time griphopiths appeared in Europe, at the very end of the early Miocene and into the middle Miocene (~17 to 11.5 million years ago), they were different from afropiths. They had molars with broader, more rounded cusps and mandibles that were low; that is, the height of the mandible corpus (the part that holds the teeth) is similar to its breadth. The cross-section is closer to being round rather than an elongated oval, the latter more typical of most early Miocene apes. Griphopiths lacked enlarged and stout incisors and canines and unlike *Afropithecus* probably were not adapted for some form of specialized feeding using the front teeth. In fact, the jaws and teeth of most griphopiths are fairly generalized and resemble those of many later-occurring hominoids, having thickly enameled molars and robust jaws (see plate 7). It is no wonder then that some fossils that we include today among the griphopiths were at one time considered to be direct ancestors of humans (which I discuss in chapter 4). Griphopiths have also been likened to the Asian sivapiths (see chapter 6). All of this is not surprising. Griphopiths are the earliest members of the great ape and human group, or clade, the hominids, and they have a morphology that is basic or primitive for that group. They share anatomy with both fossil humans and *Sivapithecus* (e.g., large molars, thick enamel). We now know, following the discovery of more complete fossils, that the human lineage, as well as that of *Sivapithecus* and the orangutan, moved on, evolving new characteristics not found in *Griphopithecus*.

As I mentioned, griphopiths first appeared in Europe around 17 to 17.5 million years ago in Germany at a site called Engelswies. Engelswies, in present day Baden-Württemberg, is rich in fossils,

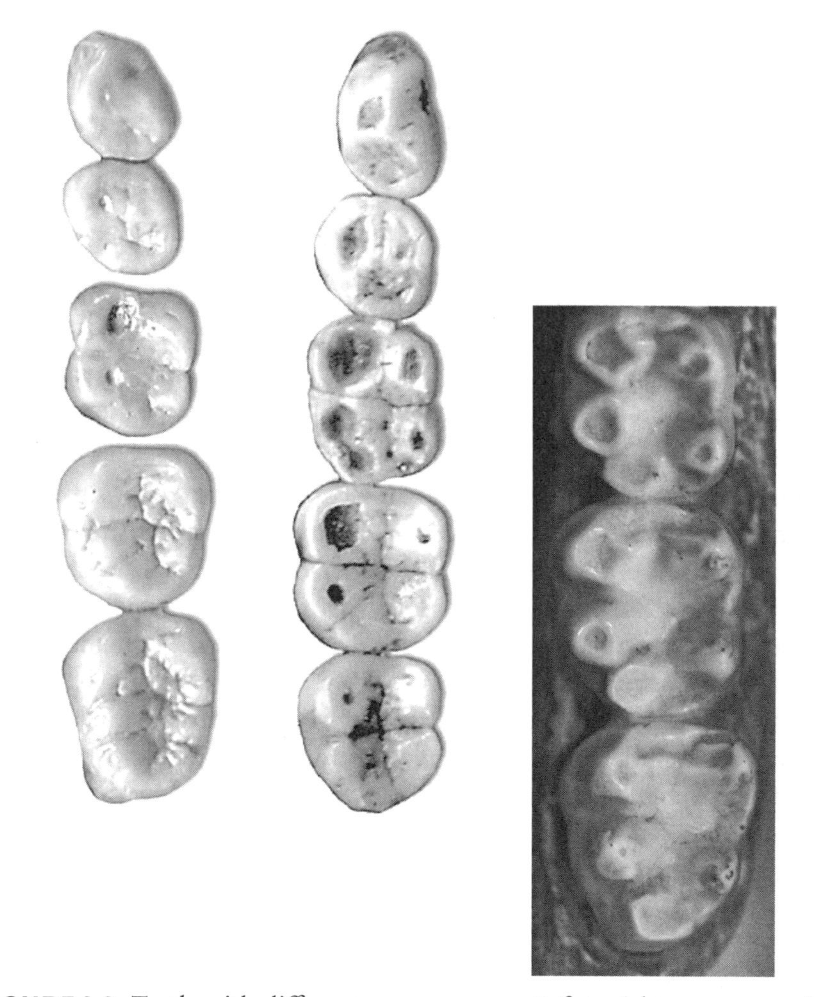

FIGURE 3.2. Teeth with different wear patterns. *Left to right*: unworn and worn molars with thick enamel (early *Homo*); unworn and worn molars of a chimp, with thin enamel and rounded cusps; unworn and worn molars of a gorilla, with thin enamel and pointed cusps. The more worn teeth are toward the top of the image, which corresponds to the front of the jaw, since these teeth erupt earlier than the back teeth. (Images by author.)

In primates with a flat EDJ (with flat dentine horns that do not project much into the enamel cap), enamel wear does not produce these isolated dentine pits surrounded by circles of sharp enamel edges; instead the tooth wears down evenly. Having a flat EDJ usually goes along with having a thick layer of enamel, which wears evenly until the EDJ is reached, at which point much larger dentine exposures appear, reflecting the broad underlying dentine horns. The consequence for the animal is the same as those having tall dentine horns: it weakens and dies.

So what does all of that have to do with the tooth from Engelswies? As I mentioned, this tooth fragment shows the wear pattern typical of a taxon with thick enamel and a flat EDJ. There are no isolated dentine pits; instead, there is very flat wear and a broad exposure of underlying dentine. Among all hominoids known from this time and before, this pattern most closely resembles what we find in the samples of griphopiths from Turkey and Slovakia. However, as noted earlier, this is a fairly broadly distributed pattern of morphology and wear, and it is possible that the Engelswies tooth fragment is not from a griphopith. It could be from an afropith, for example, as some of my colleagues have suggested. However, *Afropithecus*, while having thickly enameled teeth, had taller (less rounded), more separated or distinct cusps, and possibly a more complicated EDJ, and I do not think that *Afropithecus* teeth are as close a match for the German tooth as are many specimens of griphopith teeth from Turkey. What we really need, of course, is more fossils from Engelswies or other sites of similar age, but for now I think the idea that the half-tooth from Engelswies is the earliest evidence of griphopiths is the most likely hypothesis.

What the fossils from Engelswies tell us is that an ape with thickly enameled teeth lived there in a forest with many of the same species known from the slightly later locality of Paşalar. It is a bit surprising that no other griphopiths are known from 16 to 17 million years ago between Anatolia and Germany, but then again, fossil localities are comparatively rare in Europe. It doesn't help paleontologists that Europe is covered with vegetation and densely occupied by humans. Such areas present special challenges to the paleontologist looking for outcrops of fossil-bearing rocks—they are hiding under these geographic features.

I recently had this problem when I went prospecting for sites in the Haute-Garonne, France, with colleagues from the Natural History Museum of Toulouse. We had excellent geological and topographic maps and very good historical descriptions of the places in which fossils had been found. However, once a clay quarry that yielded some fossils is abandoned, it takes only a few years for it to become completely covered with vegetation. As it happens, disturbance plants, those that first appear when a cleared area is left to nature, are the most annoying, with various defenses in the form of barbs and other stinging organs. Even with well-covered arms and legs and some sharp cutting tools, it is almost impossible to penetrate deeply into these old quarries. My colleagues and I joked that we had arrived 150 years too late. A century and a half ago, the locals made quarries on their properties to gather sand, gravel, and clay for their houses and barns. Now they go to the nearest DIY big box store. The quarries are all overrun by nasty stinging and stabbing plants; however, we joked, with the effects of global warming, perhaps in 150 years the area will dry out and expose the sediments! Okay, perhaps it's an observation that only a paleontologist might find amusing.

FOSSIL APES IN ANATOLIA

What we mostly know about the griphopiths comes from two sites in Turkey, the aforementioned Paşalar, in Western Anatolia, and a site in central Anatolia called Çandır. At both sites we encountered for the first time hominids associated with rocks that represent a more mixed paleoecological setting. As I mentioned, many of the taxa present at Engelswies are known from the Anatolian localities, suggesting the presence of forests. In addition, a patient biologist going back in time to study the mammals of Çandır would likely see one or more of the several gliding rodents (like modern flying squirrels) known from the site, an obvious indication of the presence of fairly thick stands of trees. As at Rusinga in Kenya, the odd knuckle-walking chalicotheres, those giant, occasionally bipedal, browsing animals that are always associated with forests, were also present.

There were also grazers at both Anatolian sites, indicating the nearby occurrence of more open country or grasslands. Grazers, like cows, sheep, and most antelopes, eat widely available but fairly poor-quality grasses and have dental specializations to allow them to finely chop grass leaves and also to endure the special wear problems associated with grass eating. To protect themselves, under their outer layers grasses have tiny, extremely hard grains made of silicate (like sand) called phytoliths, which will over time wear down teeth. Grazers therefore have tall teeth, which gives them enough time to reach adulthood and reproduce a few times before their teeth wear out. They also have specializations of the gut, such as complex stomachs, for processing highly fibrous foods, but stomachs don't leave a fossil record.

Humans, of course, have found a way around the conundrum of dying from excessively worn teeth. We do things to make our teeth last longer, such as cooking food and processing food with tools, both of which cut down on tooth wear. We are also able to take care of our teeth, although modern diets high in sugars and starches can still wrcak havoc on our teeth. While comprehensive dental hygiene is a relatively recent widespread practice, there is actually evidence of toothpick use by early *Homo* as far back in time as 1.8 million years. Despite good dental hygiene, even human teeth will eventually wear out, or simply fall out, as we lose bone in our jaws. Humans, however, have learned to survive with teeth that are too worn to function properly or, in some cases, with few or no teeth at all. Ignoring replacement teeth such as bridges and dentures, which have been in use for a few thousand years at the most, humans have been surviving with few or no functional teeth for hundreds of thousands of years. A famous 250,000-year-old fossil from Zambia of a *Homo heidelbergensis*, a possible precursor of Neandertals, has a horrendous set of teeth that are mostly rotted away, yet this individual survived for years with this condition. A few Neandertals, which lived from about 300,000 to 25,000 years ago, also have jaws with most of the teeth missing due to age or injury, yet these individuals were able to survive for a significant amount of time after losing their teeth, as evidenced by the healing and bone remodeling that took place in their jaws. Bone remodeling occurs all the time as patterns of strain in bone change and bone responds, removing bone minerals where they are less needed and depositing them where they are. In the case of missing teeth, the bone that held the teeth in place simply disappears, with the proteins and minerals that composed the bone moved to some other spot in the body where they would be more useful. This process takes time and obviously you have to be alive for it to happen, so we know that these individuals outlived their teeth. How? A little help from their friends and family is the most likely explanation. This type of care, whether out of compassion or for more practical reasons (old folks can have memories of events in the past of great survival value to the group), appears to be unique to humans.

Getting back to the environment of the griphopiths, there are a number of mammal species at Çandır and Paşalar that have the tall teeth diagnostic of grazers, such as *Hypsodontus*. This early antelope had exceptionally tall teeth even for a grazer. Fossil rabbits, which also liked grass and more open spaces, are to be found in the environs of Çandır and Paşalar. So the environment in which griphopiths lived was mixed, meaning that there were wet forests, woodlands (forests with less densely spaced trees, discontinuous canopy cover, and open, sometimes grassy, spots), and more open spaces. It may be that the griphopiths were confined to the forests, or they may have been able to move about through different ecological settings. My guess, given their powerful jaws and thickly enameled teeth, is that they probably did venture away from the forest occasionally in search of food, when fruits were more difficult to come by in the forest.

Griphopithecus is best known from Paşalar, where it is represented by hundreds of teeth and a few jaw and limb fragments. Scientists don't agree about the age of the rocks at Paşalar. Research that I carried out with colleagues from France and Turkey led us to conclude that Çandır, which is very close in age to Paşalar based on faunal similarities, is about 16.5 million years old. This conclusion was determined from the fossil fauna at the site, which includes a number of taxa that are not known in rocks more recent than 16 million years (in other words, they went extinct by 16 million years ago, so if they are present at a site it must be older than that). Using magnetostratigraphic evidence, other researchers have concluded that both sites are no more than 14 million years old.

Magnetostratigraphic evidence comes in the form of patterns of remnant paleomagnetism in rocks. For reasons we don't understand, the earth's magnetic field periodically switches between normal and reversed polarities, which essentially means that sediments that are affected by the earth's magnetic field (most commonly those containing some iron) orient themselves either in the direction of the current magnetic field (normal) or in the opposite direction (reversed). Sediments accumulate to form layers of mud, silt, and clay that eventually harden to form layers of rock. As the sediments are deposited in these layers, minerals in them that are susceptible to magnetization retain their remnant magnetization, meaning that they remain oriented in the magnetic field in which they were deposited. So they essentially record the direction of the magnetic field at the time of deposition. Over many years, numerous transitions from normal to reversed magnetization have been documented, which led to the development of a giant bar-code-like chart called the geomagnetic polarity time scale (GPTS). Whenever possible, sediments with a certain paleomagnetic signal are also dated using absolute dating techniques such as argon-argon, which is based on the radioactive decay of isotopes contained within rocks. Or, if there are no suitable rocks for radiometric dating, sediments can be placed in a relative sequence based on their fauna (biostratigraphy), as long as the age of the fauna is known from other sites that have been dated. Thus, with thousands of localities and thousands of independent dates, each black bar and white bar in the GPTS, stretching back over 550 million years, has been correlated with a known absolute date (figure 3.3).

A famous episode in the use of the GPTS is called the Olduvai event, because it was first discovered at Olduvai Gorge, the famous fossil human locality discovered by Louis and Mary Leakey in the 1950s. Because Olduvai was the center of so much attention, given the presence of fossil humans, a great deal of effort was made to obtain absolute ages for the rocks there and to use the newly developed technique of paleomagnetic dating. It was discovered that at around 1.85 million years ago, there was a brief change within a reversed period to a normal period, the Olduvai event. So because of the research at Olduvai Gorge we know that anywhere around the world, if you can establish that the magnetostratigraphy of the site includes a layer with the Olduvai event, you can say that that layer is 1.85 million years old, even if you cannot directly date the rocks at your site with radiometric techniques.

The big problem is determining what very small part of the GPTS is preserved at your site. Rock deposited in normal or reversed periods do not look different from one another, nor is it possible to know only from the type of rock whether, say, the normal period you are sampling is the Olduvai event, or one just like it that happened a million years earlier. Usually paleontologists use the fauna from the site to narrow down the possibilities. If the fauna is most similar in taxonomic composition with that at a site having a known age of between 14 and 16 million years ago, for example, you would look to match up the short bar-code sequence from your site to the GPTS during that interval of time. But there are often a few possibilities, and these, along with some gaps in the paleomagnetic data, are the two big problems at Çandır. The paleomagnetic

FIGURE 3.3. Time scale for Çandır. Illustration of the combination of techniques used to determine the age of a site when it cannot be dated directly. Black-and-white columns represent the geomagnetic time scale, with the global time scale (*left*) and the one from Çandır (*right*). The numbers to the left are millions of years ago, and the number-letter combinations in the boxes are the labels given to those time periods by geochronologists. The entire time period depicted here is in the C5 chron, going from C5A at the top to C5D at the bottom, with shorter time period subdivisions within each longer time period. The Çandır section is short with several breaks (*jagged lines*). The column on the far right represents time periods that are determined by changes in marine faunas. (Modified from Begun et al. 2003.)

data suggest an age of either 13.5 or 16 million years ago, depending on how big the gaps are. There is no way to be sure, so my colleagues and I preferred to throw out the paleomagnetic data and rely on the faunal evidence. Hopefully, more direct evidence will emerge in the future on the age of these important sites. In

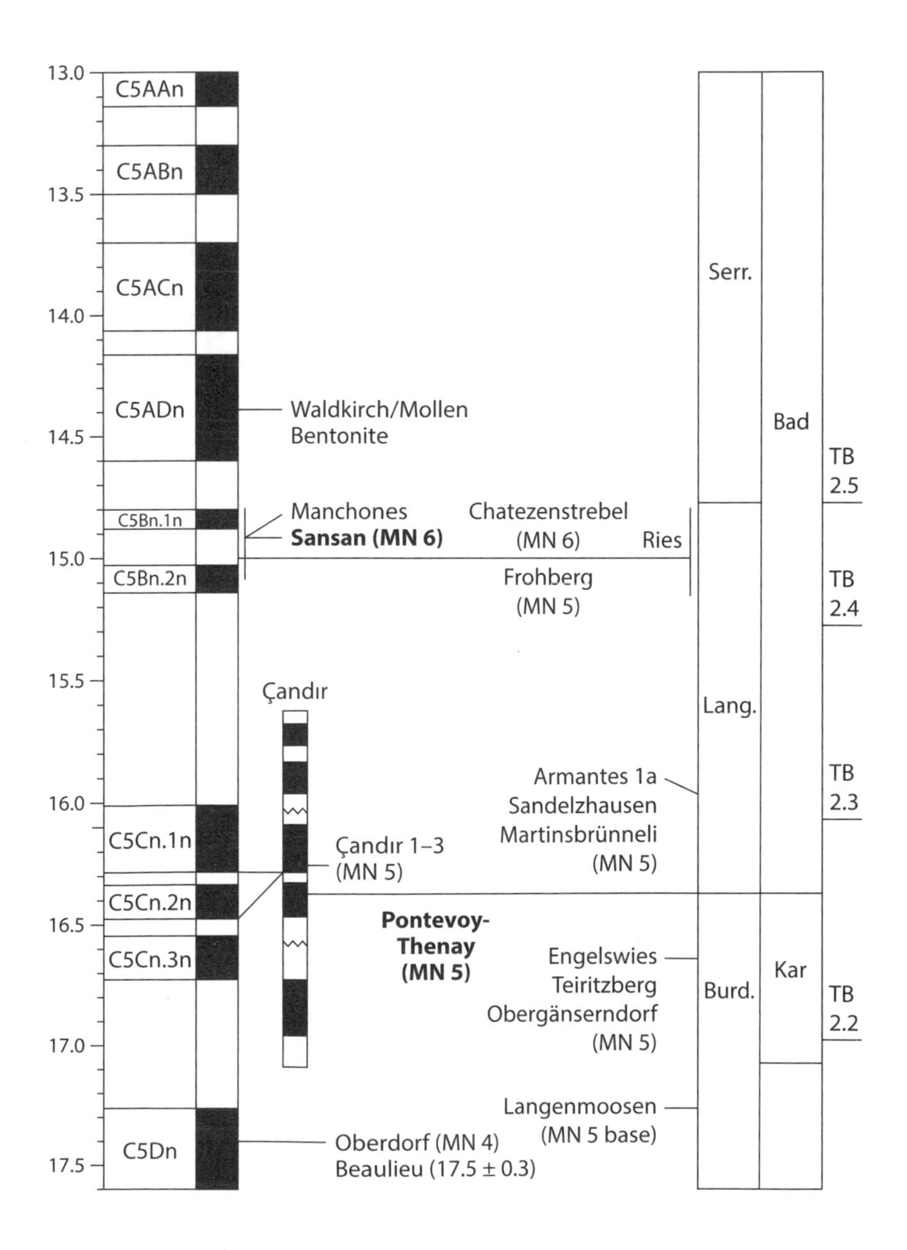

the scheme of things in this story, it actually does not matter very much, as we shall see; however, I wanted to explain some of the uncertainty in this particular case so that you can get a feel for the practical problems often encountered in determining exactly how old a fossil is (figure 3.3).

Getting back to *Griphopithecus*, our early ape with thickly enameled teeth and powerfully built jaws, we find that in details of tooth anatomy, it is more like a living great ape than is *Ekembo*. For example, *Ekembo*, *Afropithecus*, and other early Miocene apes retained well-developed cingula. These cingula are ridges of enamel on the inside, or tongue side, of the upper molars and the outside, or cheek side, of the lower ones. It is not clear why, but the cingula have been reduced over time from being very broad and shelflike in the earliest catarrhines and most earlier fossil primates, to being lost in modern apes. *Griphopithecus* was further along in this process than *Ekembo*. There are other differences in the shape and proportions of the teeth that distinguish *Griphopithecus* from older taxa.

The best-preserved jaw of *Griphopithecus* is the mandible from Çandır, the only primate specimen from the site. It is sufficiently modern looking that at one time it was touted as evidence of human origins in Turkey, but as I mentioned earlier, the resemblances to fossil humans are coincidental, having occurred independently. Other more partial jaws from Paşalar have the same morphology: that of an animal that fed on hard objects. Paleontologists haven't been able to reconstruct the face of *Griphopithecus* from the fossils recovered thus far, so it can't be compared with *Afropithecus* in much detail. But we can say that *Griphopithecus* lacked the specialized morphology of the canines and incisors that suggest that *Afropithecus* was using the front part of its jaw to pry open tough, outer coverings of fruits. As covered briefly in chapter 2 and will be seen in later chapters in more detail, the front part of the jaw is very revealing of evolutionary relationships, but unfortunately it is not well enough preserved in *Griphopithecus* for us to make any definite conclusions.

A few limb bones are known from Paşalar, the most informative of which are phalanges (finger and toe bones). They are broadly similar to those of *Ekembo* and *Afropithecus* and indicate that *Griphopithecus* moved in the trees and probably to some extent on the ground the same way *Ekembo* and living monkeys do: on the tops of branches or on the ground with their palms on the surface. Remember, this is very different from the way that modern apes move about, and we will not see the origins of the modern-ape mode of locomotion until later in this story. Like *Ekembo*, from a distance one would probably take *Griphopithecus* for a large, tailless monkey.

Griphopithecus is known from two other sites in Europe, both probably a bit younger than the sites in Anatolia. The type site (i.e., where the specimen that was used to establish the genus name was found) is actually in Slovakia, at a place called Děvínská Nová Ves, in the metropolitan region of Bratislava. Only a handful of teeth are known from Děvínská Nová Ves, which used to be known as Neudorf an der March when that part of Slovakia was part of the Austrian Empire. Some authors still refer to the site by the outdated colonial name, which does not go over well with Slovak colleagues.

The other, more informative, sample comes from a site in modern Austria called Klein Hadersdorf, which is not far from Děvínská Nová Ves.[1] There are no *Griphopithecus* teeth from Klein Hadersdorf, but there is a well-preserved humerus, or upper arm bone, and an ulna, one of the two bones of your forearm (the one on the inside of your forearm when your palms are facing forward). These bones are almost identical to their counterparts at several Africa sites (which will be discussed in chapter 4) that also have jaws and teeth closely resembling *Griphopithecus*, so I consider them to be the same kind of ape. They are similar to *Ekembo* as well and lack any adaptation for suspension seen in European apes just about 1.5 million years later.

In addition to *Griphopithecus*, another ape is known from Paşalar. For many years it was not identified as separate from *Griphopithecus* because there are only a few morphological differences. Nevertheless, a very clever analysis by my friend and colleague Jay Kelley of the Institute of Human Evolution at Arizona State University (Jay also figured out the age of first molar emergence in *Afropithecus*) revealed the presence of a second taxon. Jay noticed that among the samples of isolated *Griphopithecus* teeth there were specimens with a consistent pattern of disruption in the growth of their enamel. They had all suffered some form of stress at exactly the same time in their development, and the stress lines that had appear in their teeth as a result were all in the same place on the tooth.

Although all of these teeth had been attributed to *Griphopithecus*, the fact that they all had the same stress line suggested to Jay that they had all grown up together and had been subjected to the same stress, such as, for example, an extended period of reduced water availability—perhaps not a full-blown drought but enough of one to temporarily disrupt the normal formation of their teeth. Furthermore, the fact that all of the stress lines were found in places on the teeth that were developmentally equivalent suggests that all the individuals represented by the teeth were approximately the same age. In other words, all the individuals were hit with this stress at about the same age. This congruity suggests that the group represented by these teeth formed part of a birth cohort, the groups of animals all born during the same season in a given year. For that reason, Jay separated them from the rest of the sample and noticed subtle features that distinguished this small subsample from the majority of the specimens from Paşalar. The incisors and canines, in particular, differ from those of *Griphopithecus* and actually more closely resemble another griphopith that we will meet in chapter 4, *Kenyapithecus*. Without the evidence of a birth cohort, the differences in the incisors and canines between this small sample and the larger sample of *Griphopithecus* had been interpreted as normal variation in a single species. But the evidence that this group of apes shared a life experience that the other apes from Paşalar did not led to the identification of a second genus at the site.

So it seems that two apes lived in the vicinity of Paşalar, *Griphopithecus* and *Kenyapithecus*. This is especially interesting because *Kenyapithecus*, as the name implies, was originally found in Kenya, and the only site from which it is known in Kenya (Fort Ternan) is quite a bit younger than Paşalar (if you believe my dates). There are other taxa from Africa that also look very similar to *Griphopithecus*, meaning that there was almost certainly a series of ape range expansions between Africa and Anatolia during the middle Miocene. In chapter 4 I explore the fossils from Africa that represent evidence of these movements of animals between Africa and Europe.

LOOKING FOR FOSSIL APES IN HUNGARY

A few years ago I was invited to excavate a site in Hungary that is thought to be the same age as the younger *Griphopithecus* sites from Austria and Slovakia. It is also about the same age as the oldest sites containing the genus *Dryopithecus*, to which I will return later in this book. This site, called Felsőtárkány, is in central Hungary, not far from the Austrian and Slovakian *Griphopithecus* sites. Based on the nature of the sediments, which are river or lakeshore deposits, and the known fauna from the site, it seemed well worth the effort to begin systematic excavations at Felsőtárkány in the hopes of finding more *Griphopithecus* or perhaps *Dryopithecus*. I was particularly interested because the central European *Griphopithecus* samples are relatively poor, consisting as I said earlier only of a few teeth from one site and some fragmentary limb bones from the other.

There is a good exposure of sediments in an old trash pit directly above a thick deposit of rhyolite. Rhyolite is a volcanic lava ejected into the air when volcanoes erupt, which then settles into layers that can be very thick, and it can be dated. This rhyolite is known to be about 13 million years old. However, the most important thing about this volcanic deposit, at least to the residents of Felsőtárkány, is the fact that it is extremely thick, extensive, and easily hollowed out, which has resulted in the construction of hundreds of cellars, mostly for wine.

I was excited. Everything about this project was very promising: a good outcrop, rhyolite, a well-studied sedimentary basin, and even a primate. Years earlier a phalanx (a finger or toe bone; it is hard to tell which) from a small primate had been found somewhere in the area, although we did not know exactly where. This phalanx closely resembles the phalanges of *Epipliopithecus*, the monkey-like primitive catarrhine best known from Děvínská Nová Ves. So, although we did not know the precise location of the primate discovery, this site looked very good.

After the end of the first season, however, I was starting to have my doubts. We had done a great deal of digging, mostly to remove overburden— the sediments and dirt deposited on top of the fossil-bearing rocks. We cut through modern soil into dried lake muds from at least 10 million years ago and found a wonderful layer with spectacular leaf fossils. Finally, we reached the sediments we thought were correlated with those in which the primate had been recovered. However, after a month of backbreaking work we had found almost nothing. On top of that, because we were digging in July in an old trash pit that faces south, temperatures reached 50°C (122°F) a few times, and the trash still in the pit produced such a strong odor that we often called it a day by early afternoon. To keep our spirits up and to cool us down, there were frequent ice-cream breaks, and I even purchased a kiddie pool, which got extensive use. We had recovered a decent sample of micromammals (small rodents, insectivores, and the like—bones so small that you can usually only find them by painstakingly washing and sieving the sediments) but only a handful of large mammal fragments. And no sign of any primate.

During the off-season I studied geological maps and satellite images to see if there were other outcrops in the area. I also got hold of old photographs from the original work that had led to the discovery of the primate phalanx. It turned out that the entire hillside above the outcrop we were excavating was bare at the time of the discovery of the phalanx, thanks to the grazing of goats. In the sixty years since, a secondary forest had grown back and obscured the sediments on the hill. In season two we explored the hillside while continuing to work on the outcrop, and we did find a few spots with sediments containing fossils, but once again, we found very few large mammal fossils and no primates.

In the third season I took the unconventional step of renting a backhoe and operator to carve a trench up the hillside. I was trying to tie the spots on the hillside together with the outcrop, which was otherwise impossible given all the vegetation cover (in retrospect I should have purchased a herd of goats). We found a few more fossils and established a nice stratigraphic section that was very useful in helping us to reconstruct the paleoenvironment, but no primates. Three years was enough. I tell this story so that you know that we do not always find what we are looking for. We did everything right, but luck was not on our side at Felsőtárkány. However,

I do not regret my three seasons there. We had to look at that site to see if we could find more primates; we did not find any, but it was worthwhile in other ways. We learned a great deal about that part of the Pannonian Lake shore environment in that part of central Europe around 13 million years ago, and as I mentioned, it is close to other primate sites in Slovakia and Austria. In addition, a number of students, several of whom have gone on to careers as professional paleoanthropologists, were trained at Felsőtárkány, and I met a number of colleagues in Hungary with whom I continue to collaborate.

MIOCENE ANIMALS ON THE MOVE: PALEOBIOGEOGRAPHY

Mariam Nargolwalla, who did her PhD with me, studied the patterns of faunal exchanges between Eurasia and Africa during the Miocene for her dissertation and found that many of the exchanges between north and south that I am suggesting here for the apes actually also involved a large number of mammals of diverse size and adaptation. We still have a poor appreciation of the complexity of how animal populations move over time, a field of study that is known as paleobiogeography. It is clear that many mammals do not tend to stay in one place, and over relatively short periods of time, they can expand their ranges across tremendous distances.

A famous example is that of the horse *Hipparion* and its close relatives. *Hipparion* is the first horse of basically modern appearance and had a skull, teeth, and limbs that closely resemble those of modern equids (horses and zebras), with one major exception: it had three toes on each limb instead of one. Modern horses usually retain thin splint bones that are vestiges of these extra toes. *Hipparion* first appeared in North America around 11.5 million years ago, and in an instant of geological time, it was found throughout Eurasia and Africa, having dispersed from North America when the conditions were right. This dispersal is so obvious that we use it to identify a biostratigraphical time period in Europe, the Vallesian, during which many apes evolved as well.

During the early Miocene, elephants extended their range from Africa to Eurasia and eventually North America, and like the horses, they have given us an important marker for a time period, the Proboscidean datum. Land mammals could not move from Africa to Eurasia before about 19 to 20 million years ago because the two continents were separated by water. The Atlantic Ocean was connected to the Indian Ocean via the Mediterranean and the Tethys Seaway, covering all the land between the actual Mediterranean and the Persian Gulf. As the African plate moved north and rotated counterclockwise, it eventually pinched off this seaway, establishing both a land bridge that mammals could disperse across as well as forming the Paratethys inland seas. These inland seas were very influential in determining the distributions of mammals in the Miocene of Europe. In addition to influencing microclimates, as the North American Great Lakes do today, many ape sites are found along the ancient shores of the Paratethys or the smaller seas and lakes that developed as the inland seas broke up. The development of these shoreline habitats apparently created new niches for apes and other forest animals to fill.

Shortly after the first elephants found their way to Eurasia, the floodgates were open for many other large mammal migrants. The earliest catarrhines appeared in Asia by about 18 million years ago, and by 17 million years ago hominids had dispersed into Europe. But the highway between Africa and Eurasia was not one way. Typically modern African forms, such as giraffes, rhinos, and antelopes, expanded into Africa in the early Miocene from Eurasia. Extinct mammals such as the carnivorous amphicyonids, or bear-dogs, moved into Africa at this time as well. So it is not such a stretch to see the evolution of the hominids as having multiple stages, with complex waves of expansions and withdrawal in response to ecological conditions over the course of several million years. In chapter 4 we will see what happened when populations of apes from Eurasia returned to Africa.

CHAPTER 4

HOME AGAIN: THE NEW AFRO-EUROPEAN APES

By 17 million years ago, the griphopiths were firmly established in Europe, and they flourished. Their more modern dentition seems to have evolved at the right place and time, allowing them to spread across the circum-Mediterranean region from Germany to Kenya. By the end of the middle Miocene, there were also apes as far south as Namibia, which may have been part of this radiation, or they may have been side branches of the earlier *Afropithecus* expansion. At about 15 million years ago, 2 million years after their first appearance in Europe, griphopiths turned up in Africa (figure 4.1).

At that time, entire faunas were moving between Eurasia and Africa, and the griphopiths were part of a bigger pattern as land mammals moved in many directions around the Old World. Sites in Kenya, Austria, Germany, and Turkey share many species that appear to have expanded and contracted their ranges over several million years. The same genera or species of rodents, pigs, carnivores, and antelopes are all found across the region from Germany and Turkey to Kenya. Three different rhinos were spread out across this area, as were several pig species. I see this zone as the Grand Central Station of the middle Miocene. Arrivals and departures came and went from this region, dispersals of many mammal lineages representing the ancestors of many of the large mammals of the Old World today. It is almost surely out of this dynamic core of expanding and contracting populations that many of the tropical and subtropical mammals of the Old World originated, including, of course, apes and humans.

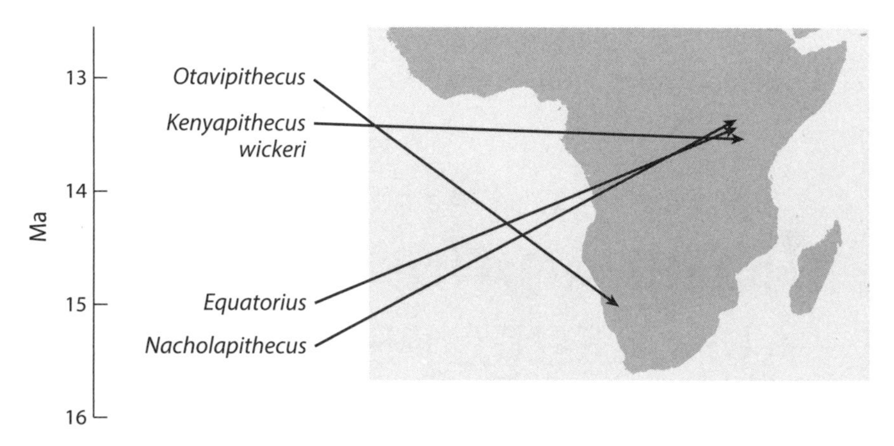

FIGURE 4.1. Timeline graphic showing time span from about 13.4 to 16 million years ago with a map showing where the fossils were recovered (Kenya, *Kenyapithecus*, *Equatorius*, and *Nacholapithecus*; and Namibia, *Otavipithecus*). Most of these sites are in Kenya, but there were probably many other places in Africa where apes lived, given the presence of an ape, *Otavipithecus*, in southern Africa. These sites are on average younger than the oldest ape sites in Europe.

ENTER *EQUATORIUS*

Equipped with their robust jaws and thickly enameled teeth, the griphopiths spread into Africa and quickly diversified into a number of new genera. *Equatorius* is known from a number of sites in Kenya and is about 15 million years old.[1] Among all the African members of the griphopith group, *Equatorius* is the least changed from its Turkish predecessors. In fact, the teeth of *Equatorius* are so similar to those of *Griphopithecus* that I have suggested that they may belong to the same genus. However, *Equatorius* has many more parts of the skeleton than *Griphopithecus*, and so it provides us with a much better glimpse of the biology of this ape.

The best collection of fossils of *Equatorius* comes from the site of Maboko Island in Lake Victoria, although the most complete specimen, a partial skeleton, was recovered from another site in Kenya called Kipsarimon. Maboko is a rare example of a site from the

Miocene in which both apes and monkeys are found. *Victoriapithecus*, which was mentioned earlier, is one of the oldest and most primitive of the Old World monkeys, and it is among the fossils more commonly found at Maboko. Recent analyses suggest that the ecology of Maboko was forest mixed with more open patches. Some have suggested that *Equatorius* may have spent more time on the ground than most other Miocene apes, but that is still uncertain. Although some researchers have drawn a connection between the possible terrestrial adaptations of *Equatorius* and terrestriality in African apes and humans, it is doubtful that there is a direct connection, given the age of *Equatorius* and the rest of its anatomy, which is considerably more primitive than that of later Miocene apes (figure 4.2).

In addition to the robust jaws and thickly enameled teeth that closely resemble those of *Griphopithecus*, *Equatorius* is also known from limb bones that show little change from the morphology in *Ekembo* and other early Miocene apes. The main differences are subtle indications of a more terrestrial adaptation in *Equatorius*. The hip joint is configured more like that of a monkey than of an ape, with muscles that were positioned in such a manner as to favor terrestrial postures. Some researchers have interpreted attributes of the anatomy of *Equatorius* to claim that this ape was a knuckle-walker, as were African apes. I doubt this, mainly because those same attributes are found in non-knuckle-walkers. In addition, *Equatorius* could not fully extend its elbow, just like *Proconsul* and most other primates but unlike all modern apes, including knuckle-walkers. The arms and hands of *Equatorius* are also similar to those of *Ekembo* and unlike those of either knuckle-walkers or highly terrestrial monkeys, like patas monkeys. It may have been more terrestrial than *Ekembo*, but it was still a forest dweller that was adept at moving about the trees. *Equatorius* lacked the specializations of both the highly arboreal and suspensory Asian apes and the specialized terrestrial African apes. *Equatorius* was a generalized ape that represents a good potential ancestor to the living great apes.

The history of *Equatorius*—or *Kenyapithecus*, *Ekembo*, and *Sivapithecus*, as it has been variously identified—is long and reveals its status as a more modern ape than *Ekembo*. Fossils from Maboko have been collected since the 1930s, and at first they were interpreted

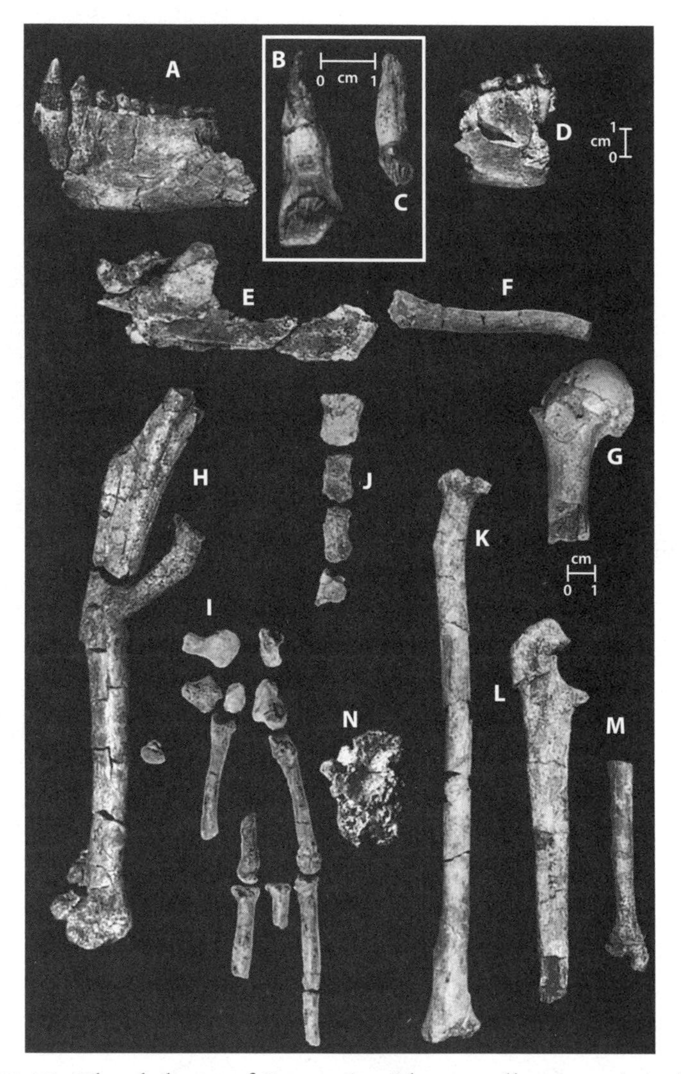

FIGURE 4.2. The skeleton of *Equatorius*. The overall pattern is quite like that of *Ekembo*, with arms and legs of equal length, a long back, and less mobile elbow joints. But the jaws are more robust, as in *Griphopithecus*. (Image courtesy of Steve Ward.)

to be *Proconsul*, now mostly referred to as *Ekembo*. In 1951, when Louis Leakey and Wilfred Le Gros Clark described a palate discovered during fossil collecting at Maboko, they did not attribute it to *Proconsul* but instead named it *Sivapithecus africanus*. The funny thing is, Leakey thought that the fossil actually came from Rusinga Island and was sympatric (living in the same place and time) with *Proconsul*, now referred to as *Ekembo*. It was not until years later that it was concluded, based on morphology and the chemical analysis of remnants of sediment still clinging to the fossil, that it probably came from Maboko, which made it about 3 million years younger than most of the fossils from Rusinga. Leakey and Le Gros Clark did not know about, or did not think relevant, the discovery years earlier of *Griphopithecus* when they named *S. africanus* but instead focused on similarities they saw with the large sample of fossil apes from South Asia (which will be our subject in chapter 6).

Leakey and Le Gros Clark focused on the large teeth—their low, rounded cusps and shallow chewing basins—of the specimen from Maboko, which indeed do resemble the teeth of *Sivapithecus*. At the time, *Sivapithecus* was being touted by some as a human ancestor, and Leakey in particular was forever looking for human ancestors.

Before going further I should say that Leakey was a tireless field researcher who discovered countless fossils, even if he exaggerated the importance of some of them a bit. He did attract the interest of many funding agencies and was instrumental in the engagement, for example, of the National Geographic Society in human-origins research, for which funding is essential. Le Gros Clark was a brilliant comparative anatomist and the dissertation professor of countless anatomists and paleoanthropologists. However, like all of us, they were not immune to error.

When *Kenyapithecus wickeri* was discovered in the early 1960s, also by Leakey, he concluded that it closely resembled australopithecines, human ancestors with large, thickly enameled teeth. By 1967 Leakey had identified additional fossils, including a mandible, this time actually from Rusinga, that he thought belonged to the same taxon,[2] and he included these and the *Sivapithecus africanus* palate in the taxon *Kenyapithecus*, recognizing a new species, *Kenyapithecus africanus*. He and other researchers at the time grouped *Kenyapithecus* with another taxon from South Asia called *Ramapithecus*, and they considered these taxa to be the earliest human ancestors; in the parlance of the time, the earliest hominids.[3]

Ramapithecus looked a lot like *Sivapithecus* but was smaller and had small canines, which was a similarity with australopithecines and humans. Reduction in size of the canine teeth is considered to be a crucial change in human evolution, so taxa with small canines are attractive candidates for human ancestry. Researchers such as Elwyn Simons had long argued for the hominid status of *Ramapithecus*.

Just about every researcher interested in the origin of the human lineage was convinced by the arguments that *Ramapithecus* and *Kenyapithecus* were the same thing and that they were excellent candidates for human ancestors. Since *Ramapithecus* was named first, both taxa were subsumed into it. *Ramapithecus* in Kenya (first *Proconsul*, then *Sivapithecus*, then *Kenyapithecus*) was at one time thought to be early Miocene in age, based on the mistaken conclusion that it was present at Rusinga, and this placed the divergence of great apes and humans at 17 million years or older. Leakey and others made much of this conclusion, because it indicated that the human lineage originated in Africa and had an extremely long evolutionary history there. This actually harkens back to ideas from the beginning of the century that emphasized the specialness of humans compared with the apes and the probability that humanness must have taken a great deal of time to develop. Leakey even attributed some apparently smashed bones and a battered-looking piece of lava found at another *Kenyapithecus* site, Fort Ternan (see below) to the deliberate actions of *Kenyapithecus*, making it a Miocene toolmaker. This interpretation has since been widely rejected.

THE MOLECULAR REVOLUTION IN PALEOANTHROPOLOGY

All of this was occurring at the same time that molecular biologists were trying to date the divergence of great apes and humans by comparing the differences in various proteins they share. In a famous study published in 1967, Vincent Sarich and Allan Wilson used immunological techniques to compare the blood proteins of African apes and humans. Sarich and Wilson calculated what they called immunological distance and, assuming a standard rate of change, estimated that the amount of difference between African apes and humans probably took about 4 to 5 million years to accumulate. This was, of course, much too late for the tastes of virtually every paleoanthropologist, and there was a great deal of pushback from that community against the validity of the molecular dating techniques. Fortunately, in recent years there has been increasing consensus between molecular biologists and morphologists on the timing of many events in evolution in general and of great ape and human evolution in particular.

On the molecular side, as described earlier, techniques have vastly improved over the past thirty years, and today, instead of blood proteins being used to indirectly assess amounts of genetic change, the actual genetic material, DNA, is. Although there is some diversity of opinion, the majority of molecular biologists and paleoanthropologists accept that chimpanzees and humans diverged from one another between 6 and 7 million years ago. So how can this be reconciled with the "hominid" status of *Ramapithecus*?

By the mid-1970s doubt began to surface about Leakey's and Simon's interpretation of *Ramapithecus* and *Kenyapithecus*. There were two key characteristics that suggested a link between these Miocene apes and humans. One was the molars, which as I mentioned earlier have thick enamel; low, rounded cusps; and broad shallow basins, as in australopithecines and modern humans. The other was the small canine, which as noted earlier, is a key characteristic of the human lineage and is even found in the earliest australopithecines. A number of young researchers at the time, including Len Greenfield and Dave Frayer, demonstrated that *Ramapithecus* is extremely similar to *Sivapithecus* and that they are in fact the same thing, *Ramapithecus* being females of *Sivapithecus*. Since *Sivapithecus* was named first, *Ramapithecus* was thus subsumed into *Sivapithecus*. This explains the small canines. *Ramapithecus* and *Kenyapithecus* do not have canine reduction as in hominins; they have small canines because all the specimens attributed to them are female. Female great apes have much smaller canines than males. The comparison between the canines of *Ramapithecus* and *Sivapithecus* is exactly the same as the comparison between the canines of male and female great apes.

But what about the humanlike molars? Many Miocene apes have thickly enameled molars with low cusps and broad, shallow basins. There are two explanations for this, and one does not necessarily exclude the other. The robust jaws and large, flat, thickly enameled teeth found in many fossil apes and *Australopithecus* may have occurred independently and are not an indication of a close evolutionary relationship. While this may seem unlikely, in fact in evolution it is extremely common for similar features to develop independently in unrelated or distantly related lineages. It is so common that the phenomenon has a technical name, homoplasy, or parallel and convergent evolution, which were described earlier. As it happens, homoplasy in the jaws and teeth is one of the most common examples of this phenomenon, especially in primates. There were only so many food sources available to primates and only so many ways to chew it, so it seems that certain tool kits for exploiting those food sources developed independently.

Another possibility is that the shared presence of robust jaws and large teeth represents the primitive condition for hominids (great apes and humans). Many Miocene apes have robust jaws with large teeth and thick enamel, and in some cases this condition was probably retained from a common ancestor with these traits. It may be, in other words, a shared primitive character, reflecting a distant common ancestor and not the supposed last common ancestor of australopithecines and *Ramapithecus*.

So the case for the hominin status of *Ramapithecus* and *Kenyapithecus* collapsed. The death blow was dealt a few years later when a spectacular discovery made by David Pilbeam's group showed that *Sivapithecus* (now including *Ramapithecus*) was probably closely related to living orangutans. Pilbeam's discovery was a fossil face with many of the highly distinctive characteristics of orangutans (*Pongo*), which will be discussed in chapter 6.

The event that really started the notion of Miocene hominin origins was the 1961 description of *Kenyapithecus wickeri*, from Fort Ternan in Kenya. The Fort Ternan site is thought to be younger than Maboko and Kipsarimon, about 13.5 million years old. Some evidence also suggests that Fort Ternan represented a somewhat more open environment than Maboko, although there was forest cover and *Kenyapithecus* was probably still dependent on trees. The critical specimen is a piece of a palate with premolars, a molar and that famously small canine. The canine was actually found detached from the palate, and when Leakey glued it back on, he placed it too high in the jaw, resulting in the appearance of a very small canine. Even when this was corrected, the canine is still small, but as noted, this is because it is from a female. Male canines are also known from Fort Ternan, but since *Kenyapithecus* was thought to be a hominin with canine reduction, these were attributed to *Proconsul*, even though there was no other evidence of *Proconsul* at the site. Today all of the ape fossils of the appropriate size and morphology from Fort Ternan are attributed to males and females of *Kenyapithecus* (see plate 8).

Like all middle Miocene apes, *Kenyapithecus* has modern-looking teeth:, thickly enameled with low cusps and broad, shallow basins. The upper molars are a bit longer than in older griphopiths, and there is no sign of a cingulum; both of these features are more like those of living great apes. Unfortunately, we don't know much more about *Kenyapithecus*. Of the 10,000-plus fossils that have been recovered from Fort Ternan, only a handful are primate and only a portion of those are *Kenyapithecus*. This makes it all the more surprising, in retrospect, that so much was made of this tiny sample of apes. There is a fragmentary mandible that is fairly typical of griphopiths, being low and robust. A very distinctive upper central incisor is known that differs from any other upper central incisor in the Miocene, except for specimens from Paşalar, which is the main reason that *Kenyapithecus* is identified from the Turkish site. The other upper front tooth, the upper lateral incisor, is also distinctive and most similar to the same tooth at Paşalar. There is no obvious functional explanation for the distinctive morphology of the incisors of *Kenyapithecus*.

In the past, researchers often felt compelled to explain morphology in purely functional terms. A highly influential paper by Steven J. Gould and Richard Lewontin called "The spandrels of San Marco and the Panglossian paradigm: A critique of the adaptationist programme" (see sidebar 4.1), is often credited as having led researchers to a greater appreciation that morphology is not always adaptive or specifically the target of natural selection. A morphological attribute can be the result of random, neutral change or as a side effect of some other target of selection.

Stephen Jay Gould and Richard Lewontin were among the most influential evolutionary biologists of the twentieth century. However, Gould in particular was prone to hyperbole and colorful historical or literary allusions. His critiques were not always fair, often painting the field with a broad brush. The spandrels of San Marco are features of the impressive dome of St. Mark's Basilica in Venice and intervene between the arches that hold up the dome. Gould and Lewontin thought they were good analogies for structures that appear as a consequence of something functional (the arches) but that are themselves not functional. Whether or not they are functional, Gould and Lewontin are not on firmer intellectual ground in concluding that they are a side effect any more than would be the conclusion that they are functional. Both are just-so stories that require some independent test to validate. As it happens, the engineer and historian of architecture Robert Marks has shown that the spandrels, while side products of the arches, also help to hold up the huge dome of St. Mark's and are therefore functional. It is not so easy to separate "functional" from "side effect." Panglossian refers to Dr. Pangloss, a character in Voltaire's famous satire *Candide*. Pangloss was the eternal and naïve optimist whose philosophy was that "we live in the best of all possible worlds," and he was a buffoon (Voltaire was caricaturing another philosopher he disliked, Leibnitz). In my view it is a bit harsh to caricaturize as "Panglossian" those evolutionary biologists with a tendency to interpret morphology as functional.

Despite Gould and Lewontin's critique of evolutionary biology in the mid-1970s, evolutionary biologists have long recognized the influence of development, body size, and structural constraint, in addition to adaptation, on many aspects of anatomy of organisms. For example, in order for a certain attribute that may have been the target of selection to grow, the entire developmental program of the organism may be affected, leading to changes in other attributes. It has been argued that selection for larger body sizes in *Homo erectus* had as a side effect an increase in brain size. Whether or not this is the best explanation for larger brains in *Homo*, the point is that changes in development can have far-reaching effects. We also know that as body mass increases, limb structure must change to accommodate larger bodies. Thicker, shorter legs in some very large animals (hippos, rhinos), for example, are not necessarily direct targets of selection but, rather, the inevitable consequence of having to support a larger mass.

Returning to *Kenyapithecus*, we find that in addition to the distinctive front teeth, which allow us to distinguish *Kenyapithecus* from other middle Miocene apes, its other unique attribute is in the area of the cheekbone, or zygomatic bone. It attaches to the maxilla, the main bone of the palate (upper jaw) and connects the middle part of the face to the upper jaw. In all early and most middle Miocene apes, this connection occurs low on the face, close to the alveolar process (the part of the jaw that holds the teeth; alveoli are the holes into which the roots of the teeth insert). In *Kenyapithecus*, the cheekbone attaches to the upper jaw above the alveolar process; that is, it is placed higher in the face. Although we have not found an actual zygomatic bone, we have enough of the process to which it attaches on the maxilla (aptly named the zygomatic process of the maxilla) to know that the zygomatic was high in *Kenyapithecus*. High cheekbones are characteristic of all modern great apes and all late Miocene great apes with one exception, *Oreopithecus* (which we will meet in chapter 8). Many researchers believe that this attribute indicates that *Kenyapithecus* is more closely related to hominids (great apes and humans) than to other middle Miocene taxa; however, it is not very much to go on. And, as with the incisor morphology, we do not have a completely convincing explanation for the functional significance of the position of the zygomatic process.

Sadly, we know almost nothing of the body of *Kenyapithecus*. All we have are a piece of the elbow joint and a few toe bones. But this lack has not prevented people from trying to put together images of what the whole body of *Kenyapithecus* may have looked like. Rather than being informed by fossils, the best known "lifelike" reconstruction of *Kenyapithecus* is inspired by the preconception that it was a direct ancestor of modern humans. It appears in the famous Time-Life lineup of human ancestors all walking toward the right to become modern humans. It is the one walking with two legs bent at the knees and a stooped back. Not only is this very unscientific, being based more on a desire to see *Kenyapithecus* as a human ancestor, somehow halfway between a quadruped and a biped, the image itself depicts an animal that would have been extremely awkward, both on the ground and in the trees. No primate moved or moves like that. *Kenyapithecus* would quickly have thrown out its back and knees in such a posture, making itself a choice target for a hungry leopard. The elbow of *Kenyapithecus* is known from the elbow end of the humerus. It is only a broken piece of the end of the bone, about 15 centimeters long, but the joint surfaces in the elbow joint are well preserved. These surfaces are much more like those in *Ekembo* and *Equatorius* than in living apes. The elbow joint is an interesting and complex joint with three bones. The humerus attaches to the two bones of the forearm, the ulna and the radius, so it needs two separate joint surfaces. The one for the ulna is called the trochlea, because it is more or less spool-shaped, depending on the species. In *Kenyapithecus* the trochlea is asymmetrical, unlike the more symmetrical trochlea of living apes. The other joint of the humerus, for the radius, is called the capitulum. It forms part of the ball-and-socket joint between the humerus and the radius that allows us to rotate our hands. Yes, that's right, your hand spins from palm up to palm down at the elbow, not the wrist. The capitulum of *Kenyapithecus* is not as spherical as that in modern apes, implying a lower range of mobility at the elbow than in modern apes, much like *Equatorius* and *Ekembo*, and suggesting that *Kenyapithecus* was not suspensory but more likely an above-branch arboreal quadruped, like living monkeys and *Ekembo*.

One interesting attribute of the humerus of *Kenyapithecus* is visible on the back of the bone. The large depression that occupies this surface in most vertebrates, the olecranon fossa, has a large joint surface in *Kenyapithecus*, which often occurs in Old World monkeys with large olecranon processes. Remember the discussion of this process in *Ekembo*. A large process, like that in *Ekembo*, *Equatorius*, Old World monkeys, and most quadrupeds provides additional leverage to the triceps muscles, which is associated with fast running, but it is not found in suspensory primates like apes. We do not know how long the olecranon process was in *Kenyapithecus* because we have only the other half of the joint, but I suspect that it would have been comparatively long, as in living monkeys. A few toe bones are known for *Kenyapithecus*. There is nothing about these bones to distinguish them from those of *Ekembo*. Chimpanzee toe bones are more strongly curved than in *Ekembo*, monkeys, and *Kenyapithecus*. While that feature may in part be related to body mass (*Kenyapithecus* is smaller than African apes), it is more likely that it is evidence that *Kenyapithecus* moved more like *Ekembo* than like modern apes.

A SPLENDID SKELETON: *NACHOLAPITHECUS*

Let us retreat back in time a bit. The last of the east African middle Miocene apes that I put in the informal category griphopith was *Nacholapithecus*. It also comes from Kenya, from the site of Nachola, which is dated to about 15 million years ago, like Maboko and Kipsarimon. Nachola is another fossil-rich locality with a diversity of primates in addition to *Nacholapithecus*, and it is ecologically similar to Maboko: it was a forest. The type specimen of *Nacholapithecus* is another partial skeleton, but this one is probably the most complete of an ape in the entire Miocene. The skull is not well preserved, consisting mostly of portions of the upper and lower jaws, but the postcranial skeleton includes most of the bones, at least from one side or the other.

The skeleton of *Nacholapithecus* is truly remarkable. I have rarely seen regions such as the shoulder and backbone preserved, along with enough of the limbs to reliably reconstruct limb proportions. I cannot emphasize enough how rare and valuable such data as these are. *Nacholapithecus* has a backbone for the most part like that of monkeys, with a long lower back, or lumbar region. As you know, apes, including hylobatids and humans, have shorter lower backs for reasons we don't completely understand. Long lower backs in monkeys, and presumably in *Ekembo* and *Nacholapithcus*, provide these

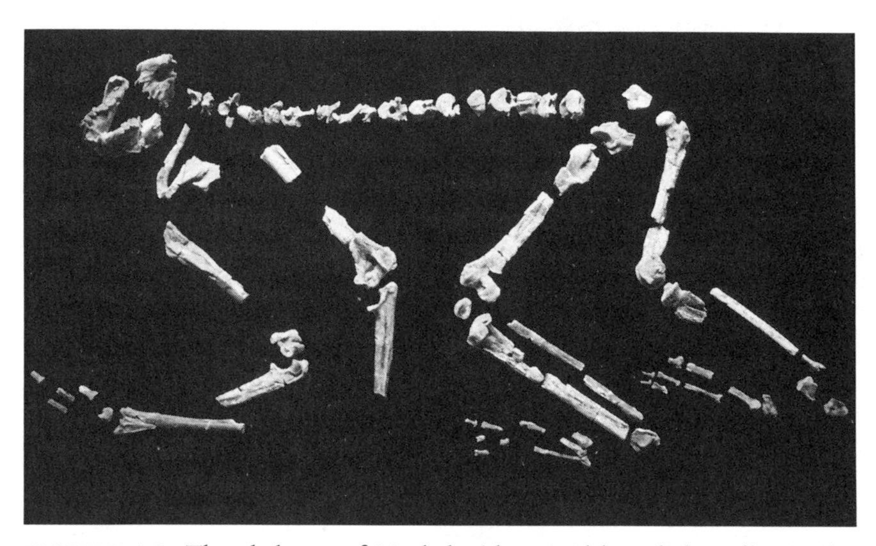

FIGURE 4.3. The skeleton of *Nacholapithecus*. Although broadly similar to *Equatorius*, *Nacholapithecus* has forelimbs with larger joints than in the hindlimbs, suggesting that it used its forelimbs in some specialized way, even though it was not orthograde or suspensory. This beautiful skeleton includes the earliest direct evidence of the presence of a coccyx. (Image courtesy of Masato Nakatsukasa.)

primates with greater limberness and flexibility. It also increases the distance between the fore and hind limbs. As I explained at the beginning of this book, these features benefit quadrupeds both in terms of agility in the trees and speed on the ground. A shorter lower back provides a greater degree of stability, which is important in animals that maintain relatively vertical postures. The loss of flexibility may be compensated for by increased mobility of the hips and shoulders in apes. At any rate, *Nacholapithecus* retains the primitive catarrhine condition (figure 4.3).

Nacholapithecus is also primitive in details of vertebra structure. In addition to having a long lumbar region, the individual lumbar vertebrae in *Nacholapithecus* are monkey-like, as in *Ekembo*, with ventrally positioned transverse processes (see plate 3). The transverse processes of a vertebra stick out to the side, as suggested by the word "transverse." Attached to them are the muscles of the back, which support the vertebral column. In most primates, the transverse processes are positioned away from the back (that is, toward the stomach, or ventral). In apes, the transverse processes are positioned closer to the back (dorsal), which changes the way the back muscles support and control movements of the backbone. Although the position of the transverse processes probably has something to do with the functioning of the lower back, the actual effects are poorly understood. It may be that the arrangement in more modern apes increases stability in the spine, while the arrangement in more monkey-like apes increases flexibility, but we really do not know. At any rate, *Nacholapithecus* is quite like most primates, including *Ekembo*, and unlike modern apes.

Most of the rib-cage vertebrae are also preserved in the skeleton of *Nacholapithecus*, as are fragments of the ribs. Together they indicate that *Nacholapithecus* had a narrow, deep chest, once again like monkeys and, in fact, most other mammals. As I explained earlier, the shape of the rib cage determines, among other things, the position of the shoulders. Primates with narrow, deep rib cages have the shoulder blades positioned on the sides of their chests, which we associate with the quadruped body plan. Quadrupeds need to support their body weight by positioning their limbs underneath their trunks. Primates, such as hominoids, with broad, shallow rib cages have their shoulder blades positioned on their backs, so that the shoulder joint faces outward. This positions the upper limbs to the side of the body rather than underneath it, resulting in a much greater range of possible positions for the arm. Paleoanthropologists associate this trait with one thing: locomotion via suspension. The fact that humans have this same shoulder configuration is one of the reasons researchers have concluded for years that humans evolved from a suspensory primate of some sort. So the number of vertebrae, their morphology, and the shape of the rib cage indicate that *Nacholapithecus* is monkey-like in trunk anatomy and probably moved like a monkey and not like a living ape.

On the other hand, the end of the vertebral column of *Nacholapithecus* is preserved, and it is a coccyx. This is the only Miocene ape with a coccyx preserved, and it is a completely unambiguous indication that *Nacholapithecus* did not have a tail, an important shared derived character of the hominoids. There is, you may recall, indirect but strong evidence that *Ekembo* also lacked a tail, and this may be related to an increase the grasping capacity of the hands. There is also evidence of this trait in *Nacholapithecus*.

The forelimb of *Nacholapithecus* is represented by many of its bones. Although it was positioned on the rib cage as in monkeys and not apes, there is an attribute of the forelimb that may foreshadow the evolution of ape suspensory characteristics: the detailed morphology of the forelimb is similar to that of *Ekembo*. For example, the humerus is configured to fit on the side of the chest, and the elbow joint has an extended olecranon process, making it incapable of full extension. The elbow joints are similar to those of *Ekembo*, *Equatorius*, and *Kenyapithecus*, and the lower end of the forearm, the wrist, is also like that of *Ekembo* and *Equatorius*, having an ulnar styloid process that articulates with the bones of the wrist. Recall that in apes the ulnar styloid is so reduced that it does not actually touch the wristbones, making it much more mobile and suitable for suspension. The wristbones themselves also resemble those of *Equatorius* and *Ekembo*, as do the finger and toe bones, having none of the characteristics found in modern hominoids that swing under branches. However, the one attribute of the forelimb that distinguishes *Nacholapithecus* from *Ekembo* is that all of the bones of the forelimb are relatively large and robust (in *Equatorius* the hindlimbs are not sufficiently preserved to be able to say how they compare with the forelimbs). Compared with the size of the teeth and especially the size of the bones of the hindlimb, the forelimb is enlarged. In the description of the skeleton, the authors, headed by Masato Nakatsukasa of Kyoto University and Yutaka Kunimatsu of Ryukoku University, both in Japan, make an interesting comparison. They show that while the joints of the forelimbs of *Nacholapithecus* are very similar in size to those of a chimpanzee, the joints of the hindlimb are noticeably smaller. *Nacholapithecus* does not share appreciably longer forelimbs than hindlimbs with living apes, but their forelimbs are comparatively robust or powerfully built. So the pattern in *Nacholapithecus* is unique, but one could argue that it is a precursor to the modern hominoid condition (see plate 9).

We do not know why the joints of the forelimb in *Nacholapithecus* are enlarged. Typically, the more stress to which joints are subjected, the larger they become, to dissipate those forces over a larger surface area, thus preventing damage. As body mass increases, joints tend to get disproportionately large. But *Nacholapithecus* isn't a big animal. In this case, it is not body mass but some type of activity pattern that must have generated higher levels of stress in the forelimb joints. It seems likely that *Nacholapithecus* was using its forelimbs to move about the forest in a different way than *Ekembo*, but also unlike modern apes, given that the forelimbs still lack any indications of suspensory adaptations. Figuring out exactly what *Nacholapithecus* was doing is tricky because there are no modern primates with the same combination of anatomical features to which we can compare it. It appears to have been unique.

Perhaps *Nacholapithecus* was a cautious climber with very powerful, grasping forelimbs that it used to steady itself above the branches. It is possible that this powerful grip allowed *Nacholapithecus* to venture onto smaller branches near the periphery of the tree to access fruits that other animals may have been unable to reach. Perhaps there was something in the way *Nacholapithecus* collected or processed food that required very powerful forelimbs. Foods that are protected by hard or tough outer coverings are generally processed with specialized features of the jaws, as may have been the case with *Afropithecus*. *Nacholapithecus* resembles *Afropithecus* in the structure of its canines, and perhaps it also used its strong hands to more efficiently extract nutrients from foods that had husks, shells, or other tough coverings. We probably will never know with certainty exactly what *Nacholapithecus* was doing with its powerful forelimbs, which is one of the puzzles that makes this ape so interesting.

Portions of the jaws of *Nacholapithecus* were also collected with the skeleton. The mandible is most similar to those of *Equatorius* and *Griphopithecus*, with a robust corpus or body, the part that holds the teeth, including thickly enameled, low-cusped molars with shallow chewing basins. The molars retain small, shelflike cingula, like in most griphopiths, but unlike in *Kenyapithecus*, which lacks cingula (although the number of *Kenyapithecus* molars recovered is very small). The lower canines are somewhat more robust than in other griphopiths, and in this way they are reminiscent of the canines of *Afropithecus*, though not as tusklike. However, it is the palate that appears most distinctive in *Nacholapithecus*.

In *Nacholapithecus* the premaxilla (the bones that hold the incisor teeth) is well preserved, though a bit crushed. As preserved, it appears to be elongated compared with that of other early and middle Miocene apes and somewhat more horizontally inclined. In cross-section (see plate 5), the premaxilla appears to overlap somewhat with the roof of the palate. This combination of characteristics distinguishes *Nacholapithecus* from all other hominoids, and as with the forelimb morphology, it is possible that it foreshadows the morphology of the hominid palate. As described earlier, in hominids the premaxilla is thick and always overlaps to some degree with the maxilla's palatine process. The premaxilla in *Nacholapithecus* is not particularly thick (from the lip surface to the tongue surface), but it does look long. It is very difficult to evaluate these attributes with confidence. As I mentioned, the area is crushed, and my examination of the original fossils leads me to conclude that the overlap between the premaxilla and the palatine process is exaggerated in this specimen. Why does this matter? Because the position of the premaxilla can offer evidence of whether *Nacholapithecus* had a more modern, apelike upper jaw. Given the preservation of the structure, a more modern configuration of the upper jaw in *Nacholapithecus* is ambiguous, and without new specimens, we will have to leave it at that.

However, I do think that overall the premaxilla in *Nacholapithecus* is more modern than that of *Ekembo*, *Afropithecus*, and *Morotopithecus*. Unfortunately, the premaxilla is not preserved in any other fossils of early or middle Miocene apes, and thus we can't compare the *Nacholapithecus* premaxilla to that of other griphopiths.

AN APE FROM NAMIBIA: *OTAVIPITHECUS*

The fourth ape from the middle Miocene of Africa is *Otavipithecus*. The most unusual aspect of this discovery is its location, Namibia. Namibia is several thousand kilometers from Kenya, and *Otavipithecus* is the only Miocene ape known from anywhere south of Kenya. The locality from which the fossils were recovered, Berg Aukas, dates to about 13 million years ago. The best specimen of *Otavipithecus* is half of a mandible. The mandibular body is fairly tall and more closely resembles that of *Ekembo* and *Afropithecus* than that of other middle Miocene apes. Most important, the teeth are very reminiscent of *Ekembo* and *Afropithecus*, especially in their size and proportions. The specimen is almost certainly a female, based on the size of the socket for the canine. The canine itself is not preserved, so comparison with *Afropithecus* is difficult, because the best-preserved mandibles of *Afropithecus* are from males.

We have a few other fossil fragments attributed to *Otavipithecus*, but they are not particularly helpful in elucidating its evolutionary relationships. The finger bone and a piece of ulna are both much like those of *Ekembo*, but then again, *Afropithecus* is essentially indistinguishable from *Ekembo* below the neck. From the biogeographic and adaptational points of view, it would not surprise me if *Otavipithecus* were related to *Afropithecus*. Unlike *Ekembo*, *Afropithecus* is almost certainly related to *Heliopithecus*, which is also found thousands of kilometers from Kenya in Saudi Arabia. The adaptations that characterized *Afropithecus* made this range extension possible, and the same may have occurred to the south.

On the other hand, descendants of *Ekembo* persist in the fossil record of Africa until about 10 million years ago, well within the time frame of *Otavipithecus*. One taxon that is thought to be a descendant of *Ekembo* is *Samburupithecus* (more on this later), which also shares dental proportions and cingula with *Ekembo* but is otherwise morphologically distinctive. Although we don't know the precise affinities of *Otavipithecus* in terms of the big picture, it doesn't really matter. *Otavipithecus* is not a griphopith and is not more closely related to hominids than are *Ekembo* or *Afropithecus*. It is a side branch, but a very interesting one. It illustrates the adaptability of early Miocene African apes and their ability to expand their ranges over vast distances. It also illustrates that fossil apes could be lurking anywhere in the fossil record of Africa. Although we focus our attention on East Africa, because that is where the vast majority of fossils have been recovered, it is quite possible that the true origin of the great ape and human lineage lies elsewhere in Africa.

So we have seen that there are four new middle Miocene apes from Africa, compared with many more in the African early Miocene and the Eurasian late Miocene. Big changes were occurring, but there was less diversity, or at least less diversity is known than in the early or late Miocene. One of these apes, *Otavipithecus*, is only known from half of a lower jaw and a few other fragmentary specimens, and it appears to be an offshoot of the *Proconsul* radiation (including *Afropithecus*), which therefore extended as far south as Namibia. The other three are from Kenya and include *Equatorius africanus*, from Maboko and Kipsarimon; *Nacholapithecus kerioi*, from Nachola, and *Kenyapithecus wickeri*, from Fort Ternan. All three are broadly similar to the European griphopiths, and I include all of them in this group, even if some of them may be more advanced and more closely related to modern apes than is *Griphopithecus*.

These apes as a group are broadly intermediate between *Ekembo* and late Miocene hominids. I call them griphopiths to emphasize their intermediate position between *Ekembo* and the hominids. However, griphopith (or, more formally, griphopithecid) is probably not a valid taxonomic category because some of these apes may be more closely related to hominids than are others. The elongated molars and the high position of the cheekbone in *Kenyapithecus* may indicate a closer relationship to hominids than to other griphopiths. If this is the case, then the griphopiths would exclude taxa (the hominids) that are actually more closely related to one of them than are the other griphopiths. This is known as a paraphyletic taxon, like the old classification of the great apes into the pongids (see chapter 1). If *Nacholapithecus* is more closely related to hominids, given the morphology of the palate or the limb proportions, then once again this would make the griphopiths paraphyletic. Paraphyly is always the problem with intermediate taxa. It is very difficult to know where to draw the lines. Are *Kenyapithecus* and *Nacholapithecus* hominids? If so, then our concept of the Hominidae would radically change, because it would include taxa that do not have suspensory morphology or, as in humans, clear indications of a suspensory ancestry. This is difficult to reconcile with the fact that hylobatids (the gibbons and siamangs), the living sister clade to the hominids, are highly suspensory, suggesting that the common ancestor of hominids and hylobatids was suspensory as well.

Of course, it is always possible that swinging beneath the branches evolved more than once in hominoid evolution, and in fact, I have suggested that this may be the case for *Morotopithecus*. Perhaps the earliest hominids were not suspensory, and suspension evolved independently in later hominids and in hylobatids. It is a classic conundrum. Do we emphasize the evidence of the jaws and teeth or the evidence of the limbs?

If I had to guess, I would say that hylobatids developed their particular form of suspensory positional behavior independently from the hominids. Below the skull, their skeleton is distinct in a number of ways, although many of these distinctions may be related to their small size. Gibbons and siamangs also have a reduced number of lumbar vertebrae (five) compared with monkeys and *Nacholapithecus*, so this reduction would have to have evolved independently as well. They have a coccyx and not a tail, but that was almost certainly also the case with *Ekembo*. Unfortunately we have virtually no fossil record of gibbons and siamangs, and until we do, we will not be able to test this or any other hypothesis of hylobatid origins. If it is the case that hylobatids became suspensory independently, then one or more of the griphopiths may in fact be very primitive hominids. But until we have better fossil evidence I prefer to keep them in the informal griphopith category.

It would certainly help if we knew more about the postcranial skeleton and skull of *Griphopithecus* and *Kenyapithecus*. Despite many years of work at Paşalar, very few postcranial fossils have been recovered. The few toe bones that are known are much like those of other middle Miocene apes. Researchers have returned to Fort Ternan over the years, but the fossil-bearing rocks are inaccessible (a rock overhang created by the excavations makes it too dangerous to work there today). We will need to find new sites with these apes to learn more about their cranial and postcranial anatomy and to clarify their relationships.

CHAPTER 5

THE BIG EAST-WEST DIVIDE

By about 13 million years ago, the fossil record of apes in Africa pretty much dries up. At the same time, the fossil record of Miocene apes picks up again in Eurasia. Within 500,000 years or so, well-preserved and clearly hominid fossils begin to appear nearly simultaneously in Europe and Asia and last until about 7 million years ago. During that time, almost nothing is known from Africa, but the few scraps that are, from Kenya and Ethiopia, are for the most part either undiagnostic—that is, we really cannot tell what they are because they are so fragmentary—or they are related to early Miocene taxa and have nothing to do with the hominids.

Thus, despite the existence of a number of localities with primate fossils, there is no solid evidence of the hominid clade in Africa before 10 million years ago. There is one locality that dates to just under 10 million years that may contain a hominid. I think it is most closely related to the European great ape *Ouranopithecus* and not to any specific lineage of extant hominid. I describe this handful of intriguing-but-difficult-to-interpret fossils in chapter 9.

At the same time, as I mentioned, the fossil record of hominids is quite abundant in Eurasia, which has led me to hypothesize that the crown hominids (lineages that include living members) actually originated in Eurasia. This hypothesis, sometimes called the "out of Africa and then back again" hypothesis (not by me), has been criticized by some as too complicated or biased given the small number of sites in Africa of the same age. I will discuss these critiques in chapter 9 as well.

At this point, it does appear that we need to look to the record in Eurasia to find evidence of the earliest crown hominids, the hominines (African apes and humans) and the pongines (orangutans). While it does not prove anything, it is comforting to know that no less a personage than Darwin was open-minded about the subject as well. In the introduction I mentioned that Darwin was receptive to the idea that European apes may be related to African apes. On the subject of ape and human evolution, the most famous quotation from Darwin's *The Descent of Man* is

> In each great region of the world the living mammals are closely related to the extinct species of the same region. It is therefore probable that Africa was formerly inhabited by extinct apes closely allied to the gorilla and chimpanzee; and as these two species are now man's nearest allies, it is somewhat more probable that our early progenitors lived on the African continent than elsewhere.

His very next line in *The Descent of Man* is more rarely remembered:

> But it is useless to speculate on this subject, for an ape nearly as large as a man, namely the *Dryopithecus* of Lartet, which was closely allied to the anthropomorphous *Hylobates*, existed in Europe during the Upper Miocene period; and since so remote a period the earth has certainly undergone many great revolutions, and there has been ample time for migration on the largest scale.[1]

I love that second quotation for a number of reasons. It is hard to overemphasize the foresight that Darwin had on the dynamics of faunal dispersals at a time when most considered that animals were confined to the places in which they are found today. Darwin was also recalling that the fossil evidence at hand cannot be summarily dismissed, despite how strongly one supports an alternative hypothesis. By "the *Dryopithecus* of Lartet," Darwin was referring to a type of great ape that we will encounter later on (chapter 8), which was considered by many to be a possible ancestor of living apes and humans. Before examining the fossil record of these lineages, it is useful to explore the context in which these apes evolved and the evidence for the dispersal of other mammals, which may make it easier to accept the proposition that stem hominids moved into Eurasia and evolved into crown hominids.

As we saw earlier, during the middle and late Miocene, vast bodies of water existed in Europe that are reduced or gone today. These were remnants of the Tethys Seaway, which at one point linked the Atlantic and Indian Oceans, cutting off any land connection between Eurasia from Africa by which land mammals could disperse. After the initial closure of the seaway around 21 million years ago, due mostly to the rotation of the African continental plate counterclockwise into Europe, the land routes between Europe and Africa periodically closed and opened over several million years. As the plates collided, new mountain ranges were created, and sea levels fluctuated, both creating barriers to animal movement. The Alps, Pyrenees, Carpathian, and Dinaric Mountains of Europe and the Pontus, Taurus, and Zagros Mountains of western Asia were all rising at this time. So the patterns of faunal exchange between Africa and Eurasia and within each of these continents, were extremely complex, with dispersals possible during certain times and impossible at other times (figure 5.1).

By about 13 million years ago, there seems to have been restricted exchange between Eurasia and Africa, especially regarding apes. The western Paratethys had disappeared. Barriers were imposed by the Rhine Grabben, a large fault in the Rhine valley that probably blocked much land-mammal movement north of the Alps between Germany and the westernmost parts of Europe and between the Alps and the Pyrenees. Despite these impediments, some mammals seem to have been able to spread across Europe at this time; we have their fossils. However, these barriers, as well as climate barriers caused by local conditions, meant that animals that reached central Europe early in time did not make it to Spain until later, which makes it very difficult to date these localities using their fauna. The sites from Spain dated by their fauna generally seem older in the middle Miocene than they probably were, because of the problems these faunas had reaching Spain. We call this discrepancy diachrony in faunal ages: faunal assemblages that look similar actually differ in age because they reached different regions at different times.

So making sense of what was happening in Europe before about 13 million years ago is difficult. Fortunately by then, exchanges between

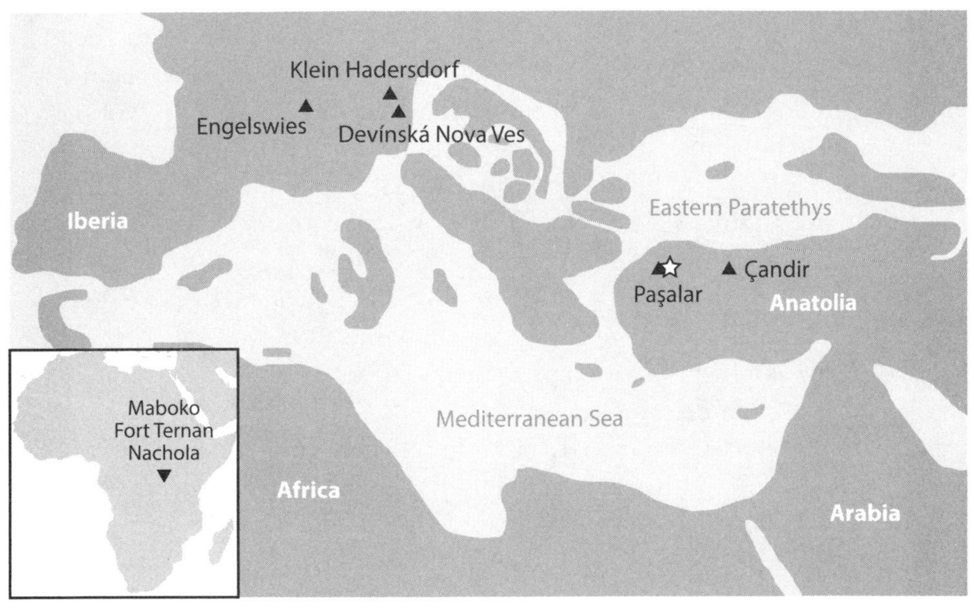

▲ *Griphopithecus* ☆ *Kenyapithecus kizili* ▼ *Equatorius, K. wickeri, Nacholapithecus*

FIGURE 5.1. Paleomap from the circum-Mediterranean region between about 13 and 16 million years ago. There is a large lake in Central and Eastern Europe connected to the Mediterranean. The most eastern portion of this huge lake system represents the paleo versions of the Black and Caspian Seas. (Modified from Nargolwalla 2009.)

central and western Europe appear to have become more straightforward. We are pretty certain that the ages of the earliest hominine sites in Spain and Europe are reliable and comparable with the ages of the earliest pongines in south Asia.

Given the presence of a very large inland sea occupying most of central Europe—the Central Paratethys (sometimes called the Pannonian Sea)—there was limited dispersal of apes between Europe and Asia and between Eurasia and Africa. In addition to the Paratethys, the mountain-building processes that had begun during the Cretaceous, upward of 80 million years ago, had led by the early to middle Miocene to the raising of mountain chains from Spain to South Asia. This huge process of mountain building, known as the Alpine orogeny, was caused, mostly, as I mentioned earlier, by the rotation counterclockwise of the African plate relative to the plates that compose Eurasia. During the time that apes were beginning to flourish in Eurasia, the terrain between the three Old World continents was incredibly dynamic. Connections were forming and closing, valleys and highlands rising and falling, and coastlines constantly changing. It is difficult to see how apes could have moved easily among Europe, Asia, and Africa between about 10 and 14 million years ago, and in fact, there is no direct evidence of ape dispersals among these regions during this time. What we see instead is the development of local ape faunas, with separate lineages arising or persisting in each region.

Other Miocene mammals experienced the same limitations as they tried to disperse. Between about 10 and 13 million years ago, there is little evidence of exchange of any land mammals between Europe and Asia, and even less with Africa. For the most part, I think a combination of the Alpine orogeny and the changing shorelines of the Central and Eastern Paratethys were responsible for this. Moving between Europe and Asia meant dispersing along a route north of the Carpathians and Alps, where there is currently no evidence of apes (although there are a few pliopithecoid sites in Poland and Switzerland). The southern route, via Turkey and the Balkans, was probably more difficult for forest-living animals, given the rough topography, though the mountains probably met the Paratethys to the north and west, and it is certainly possible that apes could have dispersed by following coastal routes. Mainly, though, we find regional development of faunas, including the apes.

By 12.5 million years ago, apes are found in Spain and South Asia that differ significantly from those of the early and middle Miocene. The morphological difference between these samples far exceeds that among the middle Miocene apes and in my view represents clear evidence of the origin of the two living clades of great apes, the Asian pongines and the African hominines. This is what biogeographers call a vicariance event, the splitting of an ancestral population into two or more daughter populations that diverge over time due to geographic isolation. In the final act of this story, we explore will the origin and evolution of each of these daughters and how they are related to living apes, including us.

CHAPTER 6

EAST SIDE STORY: OUR COUSINS *SIVAPITHECUS* AND THE ORANGUTANS

In 1837 a fossil tooth was reported from an area known as the Siwalik Hills in what was then the British Raj in India (South Asia). (In today's geopolitical reality, the Hills stretch from Pakistan in the west across northern India to Nepal in the east.) The Siwaliks have proven to be among the most fossiliferous regions of the world, and this tooth was one of the earliest to be described from there. The tooth, a canine, was likened to that of an orangutan, but the tooth has since been lost for more than a hundred years. We can never really know from what kind of animal it came; however, the fact that paleontologists at the time saw similarities between this tooth and those of living orangs is intriguing given the current consensus view that the fossil apes from South Asia have a close relationship with *Pongo*.

The history of fossil apes in Asia is complex (figure 6.1). Fossils from the region represent many taxa, and their interpretation has been controversial (some more controversial than others, as we will see with *Gigantopithecus*). I will focus here on the fossil apes that we think are most closely related to the living pongines—the orangutans.

Among the first fossil apes to be discovered were specimens from South Asia that we today attribute to the genus *Sivapithecus*. It wasn't

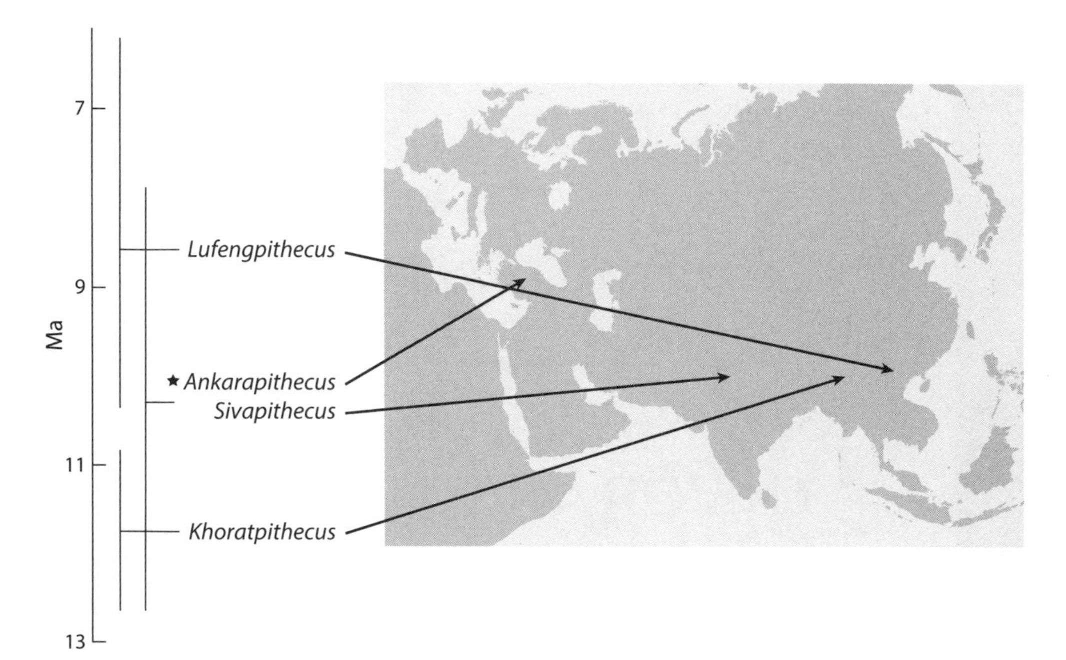

FIGURE 6.1. Timeline graphic showing time span from about 8 to 12.5 Ma with a map showing where the fossils were recovered (China, *Lufengpithecus*; India and Pakistan, *Sivapithecus;* Thailand, *Khoratpithecus;* Turkey, *Ankarapithecus*). Vertical lines denote time ranges for *Sivapithecus*, *Lufengpithecus*, and *Khoratpithecus*.

until 1879 that the first clearly diagnostic fossil ape was described from South Asia, in what is today Pakistan, and it was given the name *Paleopithecus*. This was twenty-one years after the first description of a fossil ape from Europe, *Dryopithecus* (which figures prominently in chapter 7).

By the first decade of the twentieth century, more specimens were coming to light from the Siwalik sediments, which had been deposited in a region called the Potwar Plateau of both India and Pakistan, and the specimens were beginning to attract wider interest among early paleoanthropologists. Guy Pilgrim was an important early researcher on these fossils, and he named several new taxa, including a new genus, *Sivapithecus*. Pilgrim thought that *Sivapithecus* was a hominid (which at the time meant a direct ancestor of humans, that is, hominin in our current parlance). Today all of these specimens are considered to belong to a pongine, *Sivapithecus*, for reasons that I hope to make clear.

By the 1910s, *Dryopithecus* had become a central taxon in discussions of ape and human origins. Darwin had recognized the potential importance of *Dryopithecus* forty years earlier. No fossil apes had been unearthed in Africa at the time, and *Dryopithecus* was widely regarded as an ancestral great ape, or even an ancestral African ape. It was first found in Europe (France and Germany) and later identified (incorrectly) in South Asia and Africa.

Among the first in a long line of researchers who have tried to make sense of these fossil great apes was the eminent vertebrate paleontologist William King Gregory, of the American Museum of Natural History in New York. It was Gregory and his colleague Milo Hellman who together coined the term "the Y-5 dryopithecus pattern" for the configuration of the five cusps and Y-shaped grooves that separate them that appear on the lower molars of all hominoids. This is a remarkably consistent pattern that is in fact more ancient than the hominoids themselves, since it can be found in the common ancestors of all apes and Old World monkeys 35 million years ago (see chapter 1). Gregory and Hellman's very influential description of new specimens from the Siwaliks, along with their review of the evidence from Europe, set the stage for a very lively (at times, downright nasty) debate about the significance of these fossils. By the late 1930s, many scholars were coming around to the view that all of the fossils from South Asia should be placed in a relatively small number of genera, perhaps two, as opposed to the ten or so that had been named.

Early on, Gregory was critical of Pilgrim's view that *Sivapithecus* was a hominin, but over time he would come to hold a similar view about a different Siwalik genus. By the 1930s, as the record of fossil apes and early humans in Africa grew richer, there was some skepticism about the hominin status of any ape from South Asia. One exception to this reservation was George Edward Lewis, a young researcher who led the Siwaliks-bound Yale North India Expedition. Lewis discovered new specimens, some of which he attributed to the newly named genus *Ramapithecus*, which he called a hominid (today we would say hominin.) As the decade drew to a close, Gregory, together with Hellman and Lewis, came to embrace the idea that *Ramapithecus* had a special and unique relationship with the human lineage.

Without getting bogged down in species and genus names, the taxonomic shuffling that has occurred over the years involving apes from Asia, Europe, and Africa has been nothing short of spectacular, and I do not mean that in a good way. Many new taxa were named on the basis of only one specimen, sometimes only one tooth, and pet ideas and preconceptions often colored the interpretations of researchers. Of course, this continues today and is hard to avoid, but I think we are getting better at it. There may even have been a lingering belief that Asia was a more likely place for humans to have originated, for a number of reasons.

Before 1960, mainly only australopithecines had been found in Africa, in addition to, of course, fossil apes such as *Proconsul* (now mostly *Ekembo*). In contrast, large samples of the much more modern looking and behaving *Homo erectus* were accumulating, at the time exclusively in Asia. Compared with australopithecines, *Homo erectus* has a large brain and direct evidence of complex behavior (habitation sites, controlled use of fire, stone tools of many types, and the remains of butchered animals). To many researchers, *Homo erectus* was on the cusp of humanity, or had even passed into it, whereas australopithecines were still apes. There was also an old idea associated with earlier researchers such as Henry Fairfield Osborn, also of the American Museum of Natural History, and German biologists, such as Ernst Haeckel, that Asia may have been a more suitable place for human origins, given the rigors and challenges of the climate compared with Africa.

There was more than a little racism in these views, and it is a sad fact of the history of physical anthropology that ideas about inherent racial inequality were common and thought to be supported by "scientific" evidence. Professional researchers, sometimes luminaries in the field, were responsible for museum displays such as those of the so-called Hottentot Venus. Saartjie "Sarah" Baartman (this is the Afrikaans version of her name; her actual birth name is unknown), was Khoikhoi, an indigenous group of pastoralists from South Africa. She was sold into slavery in South Africa and eventually made her way to England, where she was put on display as something perhaps more apelike than human (that is, European). To make a long story short, Baartman ended up in Paris, where she was also put on display. Upon her untimely death (she had a very difficult life after her "career" as a display item ended), her body was dissected by none other than Cuvier, and her skeleton and body cast were put on display in Paris, where they remained until 1976! Another example is that of Ota Benga, an Mbuti pygmy, who was put on exhibit in the monkey house at the Bronx Zoo in 1906. After his death by suicide in 1916, a body cast of Benga was also put on display in New York.

The worst part of this history for me is that there were many people at the time of Baartman and Benga who objected to their treatment, who saw these individuals as humans, and who were disgusted by the abuse to which they were subjected. But most of the scientific community in the nineteenth century viewed them and, for that matter, most non-Europeans as inferior. After evolutionary theory became widely accepted, they were thought by many to be living links between apes and humans. These views are completely repudiated by all legitimate researchers today, but they were influential in interpretations of ape and human evolutionary history in the not-so-distant past.[1]

Discoveries continued to be made until 1939, after which research was interrupted by World War II. Richard Dehm found a startling collection of very large apes, today attributed to the biggest species of *Sivapithecus*. In China, Gustav Heinrich Ralph von Koenigswald, then a young Dutch researcher working primarily in Indonesia (then called the Dutch East Indies), described remains of the largest ape ever found. In 1935 he named this behemoth, aptly enough, *Gigantopithecus* (at an estimated 300–500 kilograms, the largest primate that ever lived). After the war, discoveries of *Gigantopithecus* resumed in China, but it was probably Elwyn Simons's arrival at Yale University in the early 1960s that had the greatest impact on interpretations of the Asian fossil-great-ape record. I mentioned Elwyn earlier in the context of the Fayum. Elwyn is an admired (mostly; like all influential researchers, he has his detractors) elder statesman in paleoanthropology, responsible for training numerous researchers and for many discoveries, first in the Siwaliks and later in the Fayum. I looked forward to seeing Elwyn at our annual professional meetings, because it was always fascinating to talk to him about the past and present state of paleoanthropology and also because he would often have original fossils in a brief case that he would show to anyone interested.

In 1965 Simons and his student David Pilbeam published a seminal paper that included the first comprehensive revision of fossil apes since Gregory and Hellman. In this paper, they essentially dichotomized the record of fossil apes into *Dryopithecus*, which was thought to be broadly related to most great apes, and *Ramapithecus*, which they considered to be a hominid. We have covered this area in chapter 4, but it is important for understanding the history of discovery to recall how influential this idea was (see chapter 8). Simons focused most of his field efforts in Egypt but did work in the Siwaliks in the late 1960s. Pilbeam and Pilbeam's student Jay Kelley,[2] have directed numerous expeditions over the past thirty-plus years to recover additional fossils from the Siwalik sediments, moving from Simons's research areas in India to Pakistan.

David Pilbeam is one of many Simon's students who have gone on to match or exceed the status of their mentor in the field. Many of his students continue to make new discoveries and provide fascinating insights that contribute greatly to our understanding of ape and human evolution. It was Pilbeam who had helped revive Lewis's concept of the hominin status of *Ramapithecus* and who would later be the principal instrument of its demise. As I noted before, discoveries made in the late 1970s and early 1980s confirmed long-held suspicions that all the Siwalik fossil apes are pongines. It was a lively debate while it lasted, involving virtually all the luminaries of twentieth-century paleoanthropology.

Fossil pongines are now known from a wide swath of Asia stretching from Turkey to Thailand. It is an enormously larger and more impressive diversity of pongines than those remaining alive today, which are confined to a single genus and all of which live on the islands of Sumatra, in Indonesia, and Borneo, shared between Indonesia and Malaysia. The geologically oldest fossil pongines are known from the Siwaliks, but other genera are known from Turkey, Thailand, southern China, and Vietnam. While most or all of these genera date from the Miocene, they range from about 12.5 million years old to possibly as young as 6 million years. A probable relative of this awe-inspiring collection of apes, *Gigantopithecus*, is recorded in sites as late in time as 300,000 years. It is too bad that *Gigantopithecus* did not make it to the present, for at the size of a polar bear, a living example would be astonishing.

The fossil record of apes from Turkey and China were also caught up in the furor over *Ramapithecus*, and each in their turn was considered to provide evidence of the origin of the human lineage. We have reviewed the record in Turkey, involving what we today call *Kenyapithecus*. Other apes from Turkey are attributed to the hominines and the pongines, making Anatolia one of the most diverse areas for fossil apes. Researchers in the West knew relatively little about fossils recovered in China in the 1970s and 1980s, due to limited exchange of scientific discoveries during the Cold War. As in the Siwaliks, *Ramapithecus* and *Sivapithecus* were identified, as well as *Dryopithecus*. Today all three are placed by most researchers in the genus *Lufengpithecus*. While there are some who still cling to the idea that the earliest humans split from chimpanzees in Asia, most would now concede that most of the Asian apes are pongines. The status of *Lufengpithecus*, however, is unclear. Usually considered another pongine, *Lufengpithecus* in many ways resembles European apes. This example illustrates that the pongines are not easy to recognize, nor is their evolutionary history easy to reconstruct, as we shall see. So on to the evidence of pongine evolution.

SIVAPITHECUS IN SOUTH ASIA AND TURKEY?

The earliest pongines show up in the fossil record of Asia very close to the same time that the earliest hominines appear in Europe, which is another reason I think it is more likely that they originated in Europe or Western Asia than in Africa. Early *Sivapithecus* is known from Pakistan, in an area called Chinji, which consists of a number of localities fairly widespread across the landscape. The

Chinji *Sivapithecus* specimens are unfortunately more fragmentary than the specimens that led Pilbeam and colleagues to see clear affinities with living orangutans. Like all *Sivapithecus*, they have robust jaws and thickly enameled molars. They are modern in dental proportions, such as in having relatively long molars and lacking cingula and long premolars. Their incisors are large, as in living great apes, especially orangs, and their canines are relatively slender in cross-section. In these ways they resemble their counterparts in Europe more than they do the middle Miocene apes of Africa, with the exception that the European apes tend to have less robust jaws and thinner enamel.

I wonder about *Sivapithecus* in Chinji because we find another taxon called *Ankarapithecus* in Turkey, at about 10 million years ago. *Ankarapithecus* was described in the 1970s based on a very fragmentary mandible, but a nicely preserved face was recovered and described in 1980. The face suggested that it was in fact *Sivapithecus*, so the genus *Ankarapithecus* was synonymized with *Sivapithecus* (that is, it was formally decided that it is the same thing as *Sivapithecus*), which was the trend at the time. Paleontologists go through periods of lumping, when we put everything in a few taxa, and splitting, when we recognize many taxa. We have not figured out yet how to compromise between these two tendencies. When *Ankarapithecus* was described, there were many similarities noted with *Sivapithecus*. This led one of the describers of the face, the influential paleoanthropologist Peter Andrews (mentioned in chapter 2) to agree with researchers about moving away from the *Ramapithecus* hominin idea toward the view that *Ramapithecus* and *Sivapithecus* are females and males, respectively, of the pongine *Sivapithecus*. But *Ankarapithecus* is not identical to *Sivapithecus*, and it would require a new generation of researchers with fresh perspectives to discover this.

I was lucky to have been in the right place at the right time. In the mid-1990s I had the good fortune to be invited to Turkey to work on the face of *Ankarapithecus*. I had been introduced to the person controlling access to the fossils, the Turkish anthropologist Erksin Güleç, by a mutual friend, the wonderfully erudite, generous, inspirational, and accomplished paleoanthropologist F. Clark Howell,

who died in 2007 at the age of 81. I was extremely honored to have been invited by Clark to spend a semester teaching at Berkeley in 1993. Imagine picking up a ringing phone and having the voice on the other end say, "Hello, this is Bill Clinton," or "This is Stephen Hawking." That's how I remember the phone call from Clark asking me to come to Berkeley. Anyway, Erksin asked me to analyze *Ankarapithecus*, and I made arrangements to travel to Turkey.

I went to Ankara and examined the specimen, noticing that it was filled with sediment and glue in the region of the nasal cavity. I was given permission to prepare the fossil, which meant to remove the sediment and some of the glue, wax, and other materials used in the initial reconstruction of the specimen. This was an unusual situation, and I would not recommend it to novices, unless you are really sure what you are getting into.

In preparing the specimen, I discovered that the nasal cavity in *Ankarapithecus* was quite distinct from that in *Sivapithecus*, being somewhat intermediate between those of the pongines and hominines. Other aspects of the face were clearly pongine-like, and I concluded that *Ankarapithecus* was indeed its own distinct genus that preserved a more primitive morphology of the premaxilla and nasal cavity than *Sivapithecus*. Since that time, a research group led by John Kappelman from the University of Texas at Austin discovered even better preserved specimens of *Ankarapithecus* that confirm my interpretations (although John does not agree with them).

To this day I do not know how I convinced the museum director to let me disassemble her precious fossil except that, as noted, I was the right place at the right time. At any rate, this is my problem with Chinji. I know *Ankarapithecus* very well, having taken it apart and put it back together. From what I know about the Chinji sample, there is not enough preserved to say if it looks more like *Ankarapithecus* or *Sivapithecus*. Why does it matter? On one level, it does not, because they are all probably pongines of one sort or another. On another level, as I mentioned, if we could figure this out, we would be able to reconstruct a very interesting pattern of hominoid dispersals and evolution between Turkey and South Asia 10 to 12 million years ago, providing us with new possible insights into the mechanisms of macroevolution, that is, the origin of species.

Most of the specimens from the Siwaliks, including those that we can confidently assign to the pongines, range in age from about 8.5 to 10.5 million years ago. Represented are three species: *Sivapithecus indicus*, *S. sivalensis*, and *S. parvada* (mentioned earlier as the largest of the *Sivapithecus* species). Most specimens consist of jaws and teeth, but an extremely well preserved face and some very interesting postcrania tell a fascinating story about *Sivapithecus*. The most famous specimen is GSP (Geological Survey of Pakistan) 15000, a fossil that includes all the teeth of the upper and lower jaws, most of the jaws themselves, and most of the face from the palate to the top of the eye sockets and the jaw joint.

When GSP 15000 was described in 1982, it was a big shock to the paleoanthropological community. As I mentioned, there had been much discussion and unease about the *Ramapithecus* hominin hypothesis, and within the community of these researchers it may not have come as a big surprise. But to those of us just coming up in the field, and those in other areas of human evolutionary studies, it was a shock to learn all of a sudden that *Ramapithecus* and *Sivapithecus* are pongines and that the molecular biologists were probably right all along in saying that the divergence date between chimpanzees and humans was much more recent than we had believed. This is because if the South Asian fossils are not hominins, they do not represent evidence that humans diverged from other apes by 12 or 13 million years ago, thus opening up the prospect that humans and apes actually diverged much more recently (see plate 10 and figure 6.2).

There are two remarkable things about the preservation of GSP 15000. The first is how well preserved it is and how well it has been reconstructed from many, many fragments. The second thing is how well preserved it is in the right areas of the skull. The front of the upper jaw—the premaxilla and the nasal cavity—are almost perfectly preserved and show the unambiguous indicators of a uniquely orang-like upper jaw. As noted, orangs are unique in the way their premaxilla joins their maxilla, and *Sivapithecus* is basically a dead ringer for *Pongo*. The middle of the face, with the nasal aperture and the cheek bones, are quite orang-like in that the former is very tall and narrow and the latter are flared to the

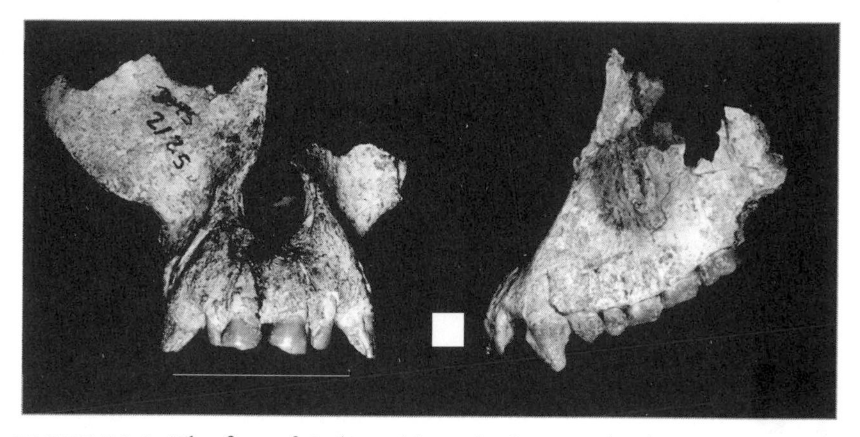

FIGURE 6.2. The face of *Ankarapithecus* in front and side views. Note the differences in facial contours and length. (Images by author.)

sides, to accommodate large chewing muscles. The upper face is perhaps the most strikingly orang-like, in that the orbits, which are so rarely preserved undistorted, are elongated ovals and not broad rectangles, as in most primates, and the space between the orbits is extremely narrow. There is no other way to describe it than to say that in the face *Sivapithecus* is strikingly orang-like. In 1982, this was quite unexpected to most of us.

The reason is that the rest of the face—the jaws and teeth—which were known from the vast majority of the collection of less complete specimens, are the least orang-like of the entire skull. So, based on what was known previously about *Sivapithecus*, the jaws and teeth showed only hints of an orangutan connection. The teeth of *Sivapithecus* resemble those of orangutans less than they do those of other Miocene apes that have thickly enameled teeth. In fact, the teeth are much more like those of other fossil hominoids having such teeth, including griphopithecids and australopiths. Compared with *Sivapithecus*, *Pongo* has molars with extremely wrinkled enamel surfaces with less distinct cusps and has less robust jaws. There were other specimens known earlier from the Siwalik sediments that show similarities with *Pongo*, which is the reason that hypothesis circulated from the earliest times, but it really took the impact of

GSP 15000 to reverse many years of conviction about the course of ape and human evolution.

If GSP 15000 were not surprising enough, limb bones described in 1989 would throw the idea of the *Sivapithecus-Pongo* connection into disarray. In that year, two limb bones, both humeri (upper arm bones), were described. These bones had a similar mosaic of features like that seen in the face of *Sivapithecus*. The lower parts of the bones, that is, the parts closest to the elbows, closely resemble those of hominoids in the way I described earlier: full extension at the elbow and a wide range of controlled mobility. However, the upper part of the bones nearer to the shoulder more closely resemble the upper arm bones of monkeys. Unfortunately, as is often the case, the most diagnostic characteristics are not present. We do not know, for example, how the humerus was attached to the shoulder joint. Was it off to the side, as in hominoids, or positioned toward the ground, as in monkeys? We simply do not know.

In my opinion, understanding the limbs of *Sivapithecus* and how they functioned requires a bit of creative thinking. The smaller bones of the limbs, wristbones of the upper limb and the bones of the foot, are not widely discussed because they are rare and small and hard to work out. I would never have looked at a single one had I not found a bunch at my site in Hungary, which forced me to learn about them. When you look at the wristbones of Miocene apes you start to see many signs of details of the evolution of great ape and human positional behavior that are not apparent from the bigger long bones of the limbs.

None of the bones of *Sivapithecus* have exactly the same morphology as those of living apes. Of course, this is exactly as predicted by evolutionary theory. These are animals that lived millions of years ago, so why should we expect their morphology to be identical to that of living animals? I recently reanalyzed two wristbones of *Sivapithecus* with my colleague and former student Tracy Kivell,[3] and we found, among other things, that these bones have functional similarities with those of chimpanzees and especially gorillas. This led us to an unexpected conclusion: *Sivapithecus*, like African apes, may have been some type of knuckle-walker. If correct, it is another interesting example of homoplasy in the positional behavior of fossil and living hominoids.

No one expected to find evidence of knuckle-walking in *Sivapithecus*, but this idea does provide an alternative explanation for the unusual forelimb morphology of this ape. Rather than seeing these bones as primitive in morphology and thus casting doubt on the closeness of the relationship between *Sivapithecus* and orangutans, this morphology may be a specialization of this fossil cousin of orangs. It may have evolved under circumstances similar to those that led to the evolution of knuckle-walking in the ancestors of African apes and humans. Given the distribution of forces and the positions of the limbs, we can explain the anatomy of the upper arm of *Sivapithecus*, as well as certain curious attributes of the knee, if *Sivapithecus* was a knuckle-walker. This reasoning does not prove that *Sivapithecus* knuckle-walked, which is, after all, an unusual form of locomotion, but in addition to explaining the other strange features of the anatomy of *Sivapithecus*, it helps us to understand how these early pongines may have spread so far across Asia. Modern orangutans are restricted to forests, where they clamber through the trees, but they are clumsy moving along the ground. Without continuous forest cover it is hard to understand how Miocene pongines could have expanded their range to the regions between Turkey and Thailand if they moved like modern orangutans. With a more eclectic form of positional behavior, such as knuckle-walking, which gives African apes efficiency on both the ground and in the trees, such a distribution across Asia seems more feasible.

LIFE HISTORY IN A TOOTH

Another fascinating aspect of the paleobiology of *Sivapithecus* is its life history. A few years ago Jay Kelley described one of those rare juvenile specimens of an individual that died just at the right time to tell us a great deal about the biology of its species. The lower jaw of this baby *Sivapithecus* contains the first permanent molar just beginning to rub against its upper counterpart. As I described earlier, if we are lucky enough to catch an individual at this stage in their development, it is possible to calculate the age at which the tooth erupted by counting the growth structures internal to the tooth and on its surface, since the tooth began to form just prior to birth.

Adding a month for prenatal growth to the total number of weekly growth bands that can be counted on this specimen, Kelley showed that this individual died at about three years of age, with its first molar just coming in. As important and touching a story of the life and death of this little *Sivapithecus* baby is, it gives us a great deal of insight into the biology of the species.

This point is important so it is worth repeating, because it is probably one of the major causes of our having evolved large brains. To recap, great apes erupt their first molars at roughly three to four years of age, whereas this tooth comes in earlier in monkeys and later (around six years) in modern humans. The fact that *Sivapithecus* shows a pattern essentially indistinguishable from that of modern great apes indicates that in all likelihood it developed at a rate and with a pattern similar to those of great apes. This being the case, we can surmise that *Sivapithecus* had a relatively large brain, grew relatively slowly, and was fairly long lived, as with all living great apes compared to monkeys. These life history characters correlate more or less with the age of eruption of the first permanent molar.

Growing more slowly, living longer, and having larger brains is generally associated with higher intelligence. Babies spend more time socializing during the time their brains are maturing, and doing so has an effect on how their brains grow and how the interconnections among the brain cells develop. Learning social roles and norms, appropriate behavior, and more complex forms of communication; developing life-long alliances and an understanding of ecological challenges; and many more cognitive advantages are enhanced in animals with extended life histories. This is almost certainly a stage through which our ancestors must have passed to reach the level of cognitive capacity characteristic of humans. Although there is more direct evidence of selection for larger brains and its associated advantages in European apes, in which partial braincases are preserved, this developmental evidence in *Sivapithecus* is important for establishing the presence of a similar phenomenon in Asia.

It is not rare to find a trend in the evolution of a lineage, such as an overall increase in size or an increase in the development of a particular adaptation (such as big chewing teeth, for example), but we do not see clear trends in the fossil evidence of *Sivapithecus*, which makes it even more difficult to understand the evolution of this genus. In the two most well known taxa, *Sivapithecus indicus* and *Sivapithecus sivalensis*, there is a slight trend toward increasing molar size from the older *S. indicus* to *S. sivalensis*, but the differences are minimal. The other widely recognized taxon, *S. parvada*, is much larger, but it is roughly intermediate in age between the other species.

Sivapithecus becomes rarer in the fossil record after about 9.5 million years ago but lasts until 8.5 million years or so ago.

ENTER THE GIANT: *GIGANTOPITHECUS*

One of the last apes from the Siwaliks is also the largest of all the Siwalik apes. Given its large size, this fossil ape is generally assigned to the genus *Gigantopithecus*. It is thought to be about 6.5 million years old, which is considerably younger than *Sivapithecus* but quite a bit older than all other *Gigantopithecus* discovered so far. There are very few specimens of this taxon, the best one being a nicely preserved mandible. While massive, it is smaller than younger *Gigantopithecus*, and it is not clear if it is really the same genus or just a very large descendant of *Sivapithecus*, which is the reason I favor the name *Indopithecus* for this late-surviving Siwalik genus. *Gigantopithecus blacki* is the other species of the genus, and it is known not from South Asia but from East Asia (China and Vietnam), from between about 1 million years ago and 300,000 years ago. It is an interesting pattern. There are few examples of lineages with earliest and latest known members separated by nearly 6 million years (figure 6.3).

This is just one of the mysteries surrounding *Gigantopithecus*. Unlike *Indopithecus*, *Gigantopithecus* is known from many specimens, but like its possible older cousin, it is only known from lower jaws and isolated teeth. This is, frankly, weird. Large animals that are represented in the fossil record by jaws and teeth are generally also known from limb bones and parts of the skull, because, being large, the bones are stronger and more resistant to destruction by the forces of nature. *Indopithecus* is at least as large as a modern male gorilla, judging by the size of its mandible and teeth. *Gigantopithecus*

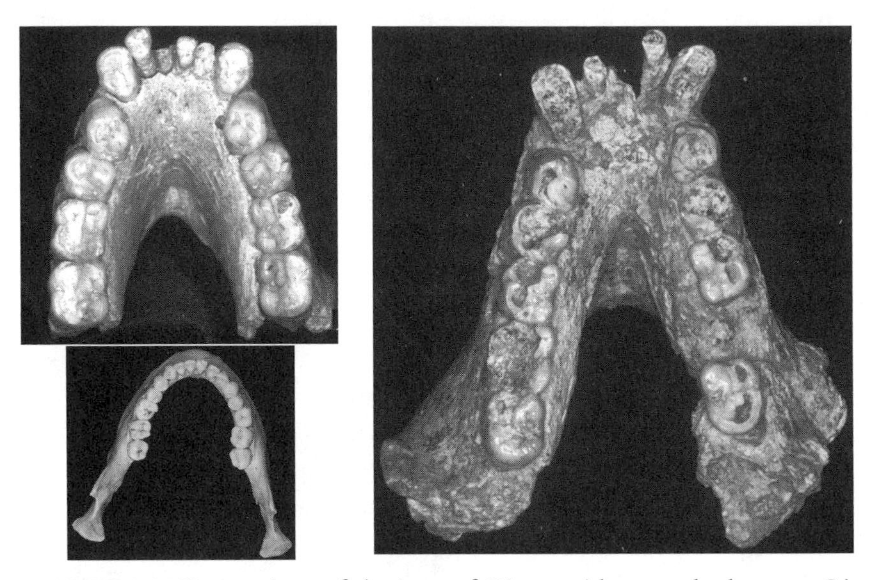

FIGURE 6.3. Comparison of the jaws of *Gigantopithecus* and a human. *Gigantopithecus* was huge. Based on the dimensions of the jaws and teeth (*top left* and *right*), it may have weighed as much as 500 kilograms, the size of a large polar bear, roughly three times bigger than a big silverback gorilla. (Top left image and right image courtesy of Milford Wolpoff. Bottom left image by author.)

appears to have been two to three times the size of a male gorilla, that is, possibly 500 kilograms, or half a ton (roughly the size of a male polar bear, the world's largest terrestrial carnivore), based on the size of its teeth. Although there are shockingly large extinct lemurs from Madagascar, one of which reached the size of a gorilla, an ape the size of a polar bear defies all expectations. Yet, despite its size and presumably its large and sturdy bones, no cranial fragments or limb bones have ever been identified for *Gigantopithecus*. This may have to do in part with the circumstances of the recovery of most *Gigantopithecus* specimens and in part, ironically enough, its large size.

The first specimen of *Gigantopithecus* described—that is, the type specimen—is an isolated tooth purchased in a Hong Kong pharmacy by G.H.R. von Koenigswald, the renowned twentieth-century paleoanthropologist we met earlier. Von Koenigswald, from the Netherlands, was working for a few years in what was then the Dutch East Indies (now Indonesia) and was aware of the presence of fossil mammal teeth in pharmacies throughout the Far East; the teeth were called dragon bones and used in the formulation of traditional medicines.[4] The exceptionally large tooth that von Koenigswald found in that Chinese pharmacy, along with numerous other teeth identifiable as *Pongo*, prompted him to eventually describe the new genus. As an ingredient in traditional medicine, dragon bones are ground into a powder, and this is perhaps the reason that larger bones and cranial fragments have not been preserved. Isolated teeth fit into little jars that fit on pharmacy shelves, but skulls and limb long bones do not. Many hundreds of Pleistocene *Pongo* teeth have also been recovered from pharmacies, and others that are known have been recovered from expeditions to various caves in China and Vietnam. There is also a skeleton of a Pleistocene orang from Vietnam recovered during an expedition. However, despite numerous attempts, no one has ever recovered limb or cranial fragments from these cave sites. Were they already scooped up and ground into power for pharmacists? Or are they so bizarre in morphology that they have not been recognized as belonging to a primate? Perhaps in the bear collections of the Institute of Vertebrate Paleontology and Paleoanthropology in Beijing there are *Gigantopithecus* bones waiting to be discovered. I would love to go through those collections one of these days.

A lot of what we think we know about *Gigantopithecus* is guesswork, given the limited fossils. Size is the first thing. Based on the size of its teeth, *Gigantopithecus* may have been up to three times the size of a gorilla, as I mentioned earlier. However, often, especially with thickly enameled species with robust jaws, which is the case for *Gigantopithecus*, the teeth are disproportionately large, that is, large relative to body mass. Even though *Gigantopithecus* had teeth that are twice the size of gorilla teeth, it is possible that it may not have been much larger than a gorilla in overall body size if its teeth were huge relative to body mass. However, it is hard for me to imagine a primate with a head the size needed to accommodate the mandible of *Gigantopithecus* on a gorilla-sized body, so my guess is that

Gigantopithecus merits its name and was in fact gigantic. The huge size of the jaws and teeth of *Gigantopithecus* strongly suggest that it was a hard-object feeder. Microwear confirms this, but researchers have also found phytoliths embedded in the enamel of *Gigantopithecus* molars.

Phytoliths are part of the strategy of grasses to avoid excessive predation. They are essentially grains of sand produced by the plant and made of silicate, and they wear down the teeth of grass eaters, or grazers. I touched on this in chapter 4. This is the reason that grazing mammals, such as most antelopes, horses, and most domesticated bovids (cattle, sheep, goats, etc.) have tall teeth.[5] They evolved tall-crowned teeth to last long enough for the animal to reproduce enough for the species to survive, despite the wear and tear of eating sandpapery grass all day long. In fact, a favorite technique used by paleoecologists to assess the ecology of a fossil site is to measure the average height of the molars of ungulates (hooved animals). Tall molars indicate dry, grassy conditions, and short molars indicate wetter conditions, favoring the growth plants for browsers (flowers, buds, fruits, and other softer, less wearing foods). Well, some *Gigantopithecus* teeth have phytoliths embedded in them, indicating that sometimes these individuals ate grassy plants (bamboo is a strong possibility).

I hesitate to bring this up, but, well, we all know about the reports of a gigantic bipedal ape roaming the forests of the northwestern wilderness of North America and parts of northern central Asia. I am talking about Bigfoot, of course. There are some who feel that *Gigantopithecus* in fact did not go extinct but evolved into or persists as Bigfoot, Yeti, the Abominable Snowman, Sasquatch, or whatever you would like to call it. I once considered, in a legitimate publication, the possibility that Bigfoot might exist as a descendant of *Gigantopithecus* but concluded in the same paragraph that this is unlikely, and I rejected all the so-called evidence of Bigfoot proposed to date. Some mysterious force draws me once again to this story. There is no evidence of Bigfoot, but I am among some (I suspect many) professionals who wish that this were not so.

Espousing even the most skeptical recognition of the possible existence of Bigfoot, without totally rejecting it, has led to many radio and TV interviews and several authors sending me their books and even Bigfoot footprint casts. Given the graininess of all the images, both photos and video, the fact that not a scrap of tissue, hard or soft, has been recovered, and the dubious authenticity of the footprints, I conclude (with some sadness) that there is no evidence of Bigfoot. But I hope to be proven wrong. I look forward to feedback from the community of believers.

FOSSIL APES FROM CHINA

The rest of the story of the fossil evidence of Asian great apes takes place in the eastern Asia. I mentioned earlier that a large collection of ape fossils from China had also been implicated in the *Ramapithecus* saga. These are the specimens from a number of sites in Yunnan Province, southeastern China.

Like South Asia, East Asia has been the target of fossil hunters for many years, first from Europe and now from China. The spectacular discoveries of astonishing diversity of the Jehol Biota in northeastern China include feathered dinosaurs and thousands of other species identified since the end of the nineteenth century. The famous American Museum of Natural History expeditions to the Gobi desert of Mongolia in the early twentieth century recovered the first dinosaur eggs known to science and numerous new species (sort of by accident; they were looking for human ancestors, which were thought to have come from Asia at the time). Some of the best preserved dinosaur fossils in the world come from the area near Lufeng, Yunnan province, southeastern China. Of course, the first large collection of fossil humans was unearthed in China in the 1920s near Beijing (at the time referred to in the west as Peking), the famous "Peking Man" site.

In sediments at Lufeng much younger than the dinosaur sites are deposits that include large samples of fossil apes. The best-known samples of fossil apes from China are from Shihuiba, commonly known as Lufeng, where hundreds of teeth, a handful of skulls, and a few limb bones have been found. Until recently, there was near consensus that the Lufeng fossil apes, appropriately named

Lufengpithecus, were closely related to *Sivapithecus* and living orangutans. A few skeptics notwithstanding, most researchers noted the very strong resemblance of the molars of *Lufengpithecus* and those of orangutans, both of which are highly crenulated; that is, their enamel surface is very finely wrinkled. There is debate on why this is the case, the most plausible explanations having to do with spreading out the forces of chewing across a number of micro ridges or perhaps slowing down the process of wear. Whatever the reason, even if it is simply a "spandrel," a random nonadaptive morphology, the teeth of *Lufengpithecus* and *Pongo* resemble each other more than any fossil ape other than *Khoratpithecus* (see below). Those of us who have looked at thousands of teeth (that's right, most of us paleoanthropological nerds have looked at *thousands* of teeth) know that crenulated teeth also occur in other apes, especially chimps, but the similarities between *Pongo* and *Lufengpithecus* in their crenulated molars are telling nonetheless (figure 6.4).

However, other aspects of the morphology of *Lufengpithecus* are decidedly un-orangutan-like. Though the skulls of *Lufengpithecus* are badly crushed, we can nevertheless tell that they lack the distinctive *Pongo* morphology of a stretched-out face and the oval eye sockets with the long, narrow space between them. The face is flat in side view, not concave as in *Pongo* and *Sivapithecus*. The front of the upper jaw is vertically oriented, not pulled out horizontally and pressed up against the floor of the nasal cavity, again as in *Pongo* and *Sivapithecus*. So, a remarkable, unique suite of features, found only in *Pongo* and *Sivapithecus*, is lacking in *Lufengpithecus*, which otherwise resembles other Late Miocene apes. So what do we make of this ape from southeast China?

From a biogeographic point of view, that is, the geography of animal distributions and origins, placing *Lufengpithecus* among the orangutans and its relatives makes a lot of sense. It is found only in Yunnan Province, China, just north of the Southeast Asian peninsula. We have fossil orangutan teeth from the Pleistocene of Southeast Asia (as well as China). and, of course, modern *Pongo* occurs only naturally in Sumatra and Borneo, islands south of the peninsula. From the evidence, we can infer that *Pongo* dispersed into the southernmost tip of Southeast Asia before the islands of Malaysia

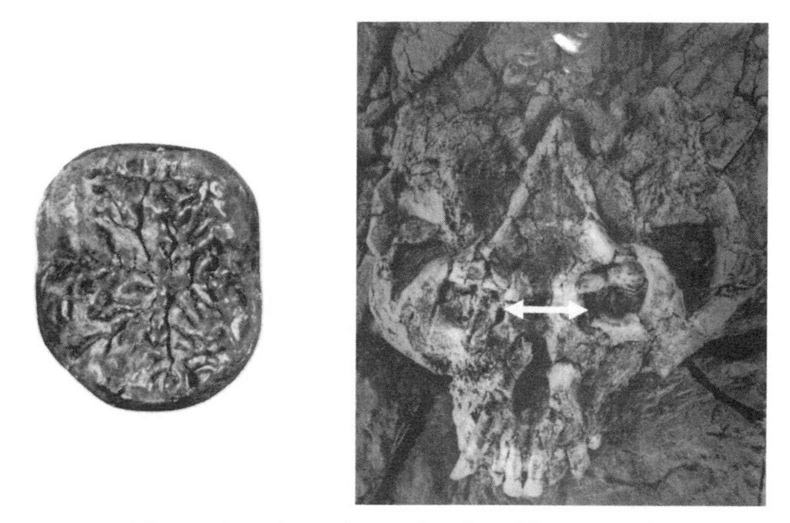

FIGURE 6.4. The teeth and cranium of *Lufengpithecus*. On the tooth, note the wrinkles (crenulations), very reminiscent of the molars of modern orangs. On the cranium, note the wide space between the orbits. Despite the distortion there is no way that *Lufengpithecus* could have resembled *Pongo* in facial morphology. (Left image courtesy of Jay Kelley. Right image by author.)

and Indonesia became separated from the mainland at the end of the last ice age.[6]

Well, if *Lufengpithecus* is not a pongine, what is it? In 2012 an analysis of the face of a very young individual of *Lufengpithecus* by published by Jay Kelley and his colleague Gao Feng. The specimen is not from Shihuiba but from Yuanmou, which is geologically older although in the same region. Kelley tentatively suggested that some or all of the sample of fossil apes from southern China may actually be related to European apes and thus to modern African apes. As I mentioned earlier, *Lufengpithecus* lacks nearly all of the orangutan-like features found in *Sivapithecus* and resembles instead African apes and European fossils apes such as *Rudapithecus* (which we will take up in chapter 7).

The problem is that all of these features shared with European and African apes are apparently primitive; that is, they may well have been present in the common ancestor of all great apes, the unusual features of the orangutan face having evolved later. So *Lufengpithecus* may be a vestige of a broad radiation of great apes without direct descendants. In addition, the specimen is a young juvenile, so it is hard to say what it would have looked like as an adult. Thus we do not know if *Lufengpithecus* does not resemble orangutans and *Sivapithecus* simply because it is primitive and has not participated in the evolutionary events that have otherwise transformed other pongines, or if it is actually a member of a different evolutionary lineage. If *Lufengpithecus* as a whole or one of the taxa from Yunnan in particular is in fact more closely related to European and African apes than to Asian great apes, this would represent a major new discovery in the ape and human-origins debate. We have to wait for more complete analyses, but for now I consider the question to be open. It could go either way.

Up until now I have mostly talked about what *Lufengpithecus* lacks that *Sivapithecus* and *Pongo* have. Let's briefly review what *Lufengpithecus* is all about. There are a few pieces of limb bones of *Lufengpithecus*. I was allowed to examine a few hand and foot bones back in the 1990s, and I concluded that the finger bones are essentially indistinguishable from those of living orangutans in the way they probably functioned. Those fingers belonged to a highly suspensory ape. That is almost certainly the ancestral condition from which all great apes and humans evolved, being present in European late Miocene apes as well. As I noted earlier, finger bones are extremely useful in deducing patterns of locomotion, especially when animals are suspensory. Based on the anatomy of its phalanges, *Lufengpithecus* was highly suspensory. Its phalanges resemble those of living orangutans in their degree of curvature and in the development of the ridges that indicate the development of the ligamentous tunnels through which the tendons of the muscles that went to the fingers passed. Each of our fingers has these tunnels on the palm side of our hands, and they are composed of ligaments that attach to each side of the finger bone. If the muscle that runs through the tunnel on its way to the tip of the finger is powerfully developed, the tunnel is strongly reinforced and leaves well-developed ridges on shaft of the bone. The effect of having strong muscles on the palmar side of the fingers and hands (flexor muscles) is to cause the finger bones to develop strong curvatures, which helps redirect forces in a way that the bone can better withstand, as well as producing a hook-like shape to the finger. *Lufengpithecus* (as well as a number of apes from Europe that we will meet in chapter 7) had this anatomy and so must have been highly suspensory, spending most of their time in the trees. Thus, while *Sivapithecus* looks much like an orangutan in its facial morphology, *Lufengpithecus* closely resembles orangutans in finger morphology—another curious and vexing puzzle for paleoanthropologists to solve.

The face of *Lufengpithecus* is more difficult to decipher. It is almost as if someone did this on purpose. There are two fairly complete fossil faces, but they look like they were run over by a truck and then put through a shredder. They are crushed in a way that is not unusual in sites with alternating layers of coal and clay, such as Shihuiba. When we look for such delicate structures as the details of the palate below the nose and the area around the eye orbits, it is a challenge to know what the original animal looked like before the specimen was crushed. In some cases, as with *Rudapithecus*, which I describe in chapter 7, technological advances such as CT scans and computer software for 3D reconstruction help when specimens are crushed, but in the case of the Lufeng fossils, the crushing and distortion is too great. Having said that, we are experienced at teasing out details of anatomy from crushed specimens in the same way that forensic examiners are, but they have DNA and the law to fall back on, whereas we have to convince our colleagues on the strength of our impressions. I am quite sure that the skulls of *Lufengpithecus*, though crushed, really lack any orangutan specializations (synapomorphies). In fact, in terms of the lower part of the face, the skulls have a "stepped subnasal fossa," a feature only found in African apes. While many colleagues would view this as a retained primitive character—that is, both orangutans and *Sivapithecus* evolved from this state—I am not so sure. In my view, the pongine condition of a smooth as opposed to stepped subnasal floor probably comes from a middle Miocene ancestor, such as *Nacholapithecus* (chapter 4); that is, pongines did not pass through a phase in which their ancestors had an African-ape-like subnasal pattern. This lends support to the view that *Lufengpithecus* is actually more closely related to African apes than to orangutans, but it definitely needs further study and reconstruction, given the way the fossils are preserved.

KHORATPITHECUS FROM THAILAND

More recently another type of fossil great ape has been described from deposits in Asia. *Khoratpithecus* is found in Thailand and dates to about 10 million years ago. *Khoratpithecus* is known from much less material than *Lufengpithecus* and consists of a handful of isolated teeth and one well-preserved mandible. The molars of *Khoratpithecus* tend to more closely resemble those of *Pongo* than do those of *Sivapithecus* in their greater degree of crenulation, as in *Lufengpithecus*. Most tellingly, along with *Pongo*, *Khoratpithecus* is the only great ape that lacks a muscle known as the anterior digastric, which attaches to the bottom of the front of the mandible, where it leaves a mark when present. Apart from the anterior digastric, if *Lufengpithecus* were only known from lower jaws and isolated teeth it would probably appear to be as orangutan-like as *Khoratpithecus*. So we have another ape with features that recall modern orangutans but in a manner not exactly the same as in *Sivapithecus* or *Lufengpithecus*.

The Miocene record of ape evolution never ceases to amaze me, and I expect that new work at these sites in southern China and Thailand will help to resolve these questions. However, the big picture remains unchanged. Apes that look more like hominines (African apes and humans) first appear in Europe and possibly Asia before we find them in Africa, leading to the hypothesis that the hominines first evolved in Eurasia and later dispersed into Africa to become the modern lineages of the African apes and humans, as we'll see in chapter 7.

CHAPTER 7
WEST SIDE STORY: THE AFRICAN APES OF EUROPE

Luck and Hard Work

Summer in Sabadell, Catalonia, northern Spain, is hot and muggy, despite the proximity of the sea and the mountains; 1991 was no exception. The struggling air conditioner in the lab of my colleague Salvador Moyà-Solà seemed to be generating more heat than cool air, and my six-year-old son André was reaching his breaking point. So we decided to go visit the site, Can Llobateres, a few kilometers out of town.

I parked on the side of the road and we got out to have a look around. We had to cross a flat, barren area between the road and the outcrops where the fossils are found, which André did with considerable hesitation, because it was used by local people as a garbage dump. With some coaxing from me, we made it to the section, a small cliff that shows the layers of rock in which we find fossils, and I explained to him what we thought was going on there. I tried to explain that this dry, treeless, grassy area with fields of corn on one side and trash on the other used to be a lush subtropical forest with a nearby river, palm trees, and a lot of animals such as rhinos, flying squirrels, zebras, antelopes, and of course apes. André was a little bored and wanted to play in the dirt. So did I. We had fun poking around for a while and moved along the outcrop until we got to a higher level that I had been meaning to have a look at for some time.

It was a second outcrop that had been carved out of the landscape when a small dirt road was built to allow farmers easier access to their fields. The exposed sediments looked similar to those in which I knew fossils could be found, but these were above the ones we had excavated the previous year. As we do, I hacked away at the section with my geologic hammer to see what the fresh sediment looked like, and it was the nice light blue-green mudstone that we like to see: not too weathered, oxidized, or reddish and thus more likely to contain fossils. I found a few fragments of snail shells, which was also a good sign that animal remains were preserved there, so I decided to show my colleague Salvador the spot the next day to convince him that we should put an exploratory pit there, in addition to our main excavation in the lower levels.

The next day Salvador and I went to the spot I roadcut. We both raised our hammers and hit the wall simultaneously in the same place I had cleaned the day before. The next thing that happened I remember in slow motion. Somehow we had hit the wall in the exact spot where a skull lay, and we watched as a tooth popped out of the wall and rolled down onto the road. We looked at each other in astonishment, both recognizing that before the project even started we had found something far more important than anything we had recovered in the lower part of the site that we thought was the more fossiliferous in four weeks of backbreaking work in 45° Celsius (113°F) heat the year before. We had planned only to do a small amount of sampling in the upper section of the site, but after that tooth rolled out, we spent the rest of the field season up there.

By the end of the month in the upper levels of Can Llobateres, we had recovered a nearly complete face of the ape *Hispanopithecus*. It was by far the best-preserved ape specimen ever found in the more than fifty-year history of research at the site. In another spot in the same roadcut, we found part of an ankle joint, and in the following year, the colleagues from Spain (I had since moved on to another project) found many more bones of the same skeleton. This is the only ape skeleton that has ever been found at Can Llobateres.

A few years later I was setting things up for my project in Hungary, trying to figure out where we would focus our efforts. This site, called R II Rudabánya, is more like an archeological excavation.

It is a fairly flat surface covered by a roof (plate 11). Every field season we have to mark off the site into a grid with posts and string in 1-by-1-meter squares, which takes a little time.

The retired geologist and long-time local friend of the project Gábor Hernyák had shown up in the morning and was ready to break some rocks, having little patience with the niceties of documentation.[1] He was eager to get to work, whacking away at rocks to find fossils, so I put him to work where he could do no harm. I asked him to clean an area of the exposed rocks that I knew from previous years was unlikely to have many fossils. While Gábor was excellent at finding fossils, we wanted to recover them in situ, that is, as they were buried, so we could understand how they ended up where we found them. Again, it is like a crime scene, where the body might be the focus but everything around it is important as well. You don't move the body until the CSI experts are finished processing the scene. So I asked Gábor to "clean" an area, by which I meant sweep the loose dirt off the solid dirt to see what the rocks look like. As I walked up the slope to my car to fetch something, I don't remember what, I heard Gábor crying "Anna, Anna, Anna." He had swept off some loose dirt to reveal the side of a jaw of a small ape, which he thought initially was the smaller of the two primates from Rudabánya, *Anapithecus*, hence the cries of "Anna, Anna, Anna."

It was embedded just below the top of the ledge of rock that we used to sit on during our lunch breaks the previous year. I remember a student wondering out loud whose derriere may have rubbed the spot to expose the fossil. At any rate, it sat there unnoticed for a year. By the end of the day I had worked the specimen out of the rock. It turned out not to be Anna, but rather Rudy, that is *Rudapithecus*, the fossil great ape from the site. This skull, which we now call RUD 200, or more affectionately, Gabi (the female version of Gábor), is the most complete one ever found of a European great ape and it dramatically affected my interpretations of great ape evolution (plate 12).

These two stories, from Spain and Hungary are my favorite personal stories about finding fossils, because they illustrate both the preparation and luck needed to find fossils. Finding fossils is indeed a combination of extensive background knowledge and preparation, endless hours applying for grants, organizing the projects, and purchasing equipment to get the job done. Once we get to the site, it's a combination of hard work in the field and pure luck.

THE DRYOPITHECINS

The apes found at Can Llobateres and Rudabánya are similar in many ways. They are among the earliest members of our evolutionary family, the Homininae, or hominines, and they are descended from *Dryopithecus*. Just a reminder: hominines include the African apes and humans of today and all the fossil relatives more closely related to living hominines than to living pongines (orangutans). The dryopithecins, a hominine tribe that includes most of the apes from Europe, first arrived there about 12 to 13 million years ago. At the time, most of Europe was in the subtropical climate zone, meaning that the climate was much less seasonal than today. The difference between winter and summer was limited, and, although there was probably a rainy and a dry season, they were also mild. Apes today are confined to the tropics, but in the Miocene they ranged into the subtropics, into regions similar in climate to Florida and the American Gulf Coast. Most of Europe is north of the subtropics today, but 12 million years ago the subtropics reached as far north as the English Channel.

A team of researchers exploring the forests in which dryopithecins lived would describe many of them as dense—having multiple canopies, or layers of branches—composed of trees that thrived in warm, humid conditions. Some localities resembled modern cypress swamps and their surrounding forests, and others, somewhat drier forests through which rivers flowed. Palms, cypress, willow, and other water-loving trees abounded near the rivers, lakes, and brackish seas, with pine and oak in the forests surrounding the area. The winters were mild but probably drier and cooler than the hot, humid summer. The forests were very different from the hot, dry climate of Mediterranean forests of southern Europe today. There the trees have thick, waxy leaves adapted to retain moisture.

Many animals that live today in the subtropics were common in the forests of Europe during the middle and late Miocene. In addition to finding primates, if zoologists could go back in time and study the ecology of these areas, they would encounter flying squirrels the size of small house cats, many species of ground and tree squirrels, and a large number of insectivores, such as shrews, moles, and hedgehogs, but in much larger numbers and wider diversity than we see in Europe today. The biologists would find beavers living in swamps, smaller than today's beavers, and when the scientists ventured some distance from the forest, they would be able to trap hamsters and dormice. They would also find a rich array of forest mammals, such as forest pigs, primitive deer, and tragulids (which we met in the early Miocene) and early antelope-like animals, perhaps most like today's waterbucks of Africa. Biologists interested in tracking the behavior of the unusual rhinos would tag them and follow their ranging patterns. These rhinos lacked horns, and some species had very short legs to facilitate moving around in the forest, intriguing our time-traveling biologists and perhaps reminding them of hippos. And the scientists would of course marvel at the gomphotherids, the elephant-sized ancestors of Pleistocene mastodons that were probably able to exploit both forest and more open environments, as do the forest elephants of Africa today.

Competing in size was *Deinotherium*, also commonly found in the forests of Miocene Europe and an extinct relative of elephants. It would have had bizarre backwardly curving tusks and giant teeth that resembled those of a tapir. Small herds of zebra-like animals, *Hipparion*, the three-toed horse, probably spent most of their time on the margins of the forest but came to the river or lake for water and grazing. Their feet with three toes, as opposed to modern horses and zebras with only one toe, might have allowed them to walk more easily in the soft terrain surrounding water courses.

Among the more discrete inhabitants of these forests our intrepid zoologists would find a very large diversity of weasel-like carnivores, or mustelids, including otters, ferrets, weasels, and larger badger-like animals. The bigger meat eaters would include the occasional hyena, but different from hyenas today, one that lacked the large and powerful jaws of living hyenas and was possibly tree climbing, a big threat to the forest primates. Although fading into the geological record, zoologists might spot an amphicyonid or bear-dog, that early carnivore we have already encountered in Africa, with no living relatives and probably filling a niche similar to today's black bears. Also rare but present were saber-toothed cats and proper cats, or felids, the ancestors of lions, tigers, and house cats of today and small early true bears intermediate in size between a large raccoon and a modern black bear. A very lucky zoologist might spot the spectacular *Chalicotherium*, which had continued to grace the forests of the Old World since the early Miocene.

Finally, zoologists exploring the forests of Europe during the Miocene would run into the occasional primate, although it would be hard to spot because primates were not numerous and they lived high in the canopy. In most of these forests the researchers would find early great apes smaller than modern chimpanzees but having similar habits, eating a wide variety of fruits and other forest foods, living in social groups, and caring for their highly dependent, slowly developing offspring. Our scientists would note that these apes rarely descended to the ground and were very adept at swinging below the branches, at bridging gaps between trees with their very flexible limbs, and at leaping from tree to tree when necessary. Sometimes the zoologists would also come across a pliopithecoid such as *Anapithecus*, a more distant relative of ours that lived before Old World monkey and apes branched off from one another. They would find pliopithecoids high in the trees walking on the tops of branches and occasionally swinging below them in search of food among the small branches at the periphery.

DRYOPITHECUS AND OTHERS

The oldest great ape from Europe is *Dryopithecus*, the oak forest ape. It was described in 1856 by Edouard Lartet, an eminent French paleontologist, three years before Darwin published *On the Origin of Species*. Earlier I mentioned the fact that Darwin referred to the dryopithecus of Lartet as a potential ancestor of the African apes and humans. Lartet was a student of Georges Cuvier, or, *Le Grand Cuvier* as he is known in France, widely regarded as the father of vertebrate paleontology. Cuvier famously rejected the idea that fossil humans existed and spent a great deal of time and energy debunking evidence that seemed to support the existence of fossil humans. In many cases Cuvier was correct, the evidence of the co-occurrence of human bones with fossil animals being unconvincing, but Cuvier went too far, probably in part as a consequence of his religious convictions, and rejected solid evidence as well.

To his credit, Lartet was more clear minded than his great teacher on this question, and he went on to make major discoveries elucidating the fossil history of both apes and humans. Twenty years before describing *Dryopithecus*, Lartet had published the first description of *Pliopithecus*, the ancestor of *Anapithecus* but which he thought was a precursor to the living gibbons. Decades later, in the 1860s, Edouard Lartet and his son Louis would be influential in the discovery and analysis of Cro-Magnon, one of the oldest fully modern humans from Europe.

The fossils on which Lartet based his *Dryopithecus* were sent to him by a Monsieur Fontan, who had received them from a worker at a clay deposit at Valentine, near St. Gaudens, France, that was mined for making bricks and roof tiles. It was not a scientific excavation in the sense we think of it today, so only a few other animals were recovered with the precious ape specimens. Lartet thought that *Dryopithecus*, named in part for the oak leaf fossils that were found at a nearby site (*dryos* is transliterated from δρυός, the Greek for "oak"), was similar to a chimpanzee in habits, and although his discovery preceded the publication of Darwin's *Origin*, he did speculate on the possible relationship that *Dryopithecus* could have with apes and humans. Darwin acknowledged Lartet's hypothesis with his comment in *The Descent of Man* (1871) that African apes could be descended from *Dryopithecus*, an opinion that has been almost completely forgotten but to which we shall return. Lartet's specimens consisted of a mandible and a humerus. Edouard Lartet astutely surmised from these fragments that the animal in question was chimpanzee-like in jaw and limb structure and probably behaved in many ways like a living chimpanzee. Today, with many times more fossils of *Dryopithecus* and its close relatives, we are able to greatly refine our reconstruction of the behavior of these fossil apes, but the reality is that Lartet was not very far off from what we think we know today.

We now know that *Dryopithecus* lived about 12 to 12.5 million years ago in the lush subtropical forests of France, in the foothills of the Pyrenees near the city of St. Gaudens and north of the Alps near Lyon.[2] It is also found in northwestern Spain, Catalonia, near Barcelona, in a protected basin called the Vallés Penédés, which is nestled between the Pyrenees and the Mediterranean. It is probably also known from the lush forested basins that existed in Styria, in southeastern Austria, before the Alps rose to their present heights. The most well studied sites in which we find *Dryopithecus* show that it lived in dense subtropical forests and that it must have spent nearly all of its time in the trees. It was a very adept climber and was able to swing below branches, as do all apes today. As I mentioned earlier, this ability is crucial to understanding the survival and success of these early great apes, because it allowed them to be much more secure in the trees, to climb to greater elevations, and to spread their body weight across a number of peripheral branches in search of food that most other large animals could not attain.

There is nothing quite like a *Dryopithecus* living today. I imagine it to be a kind of combination of orangutan, in terms of its skill in the trees; chimpanzee, in terms of its diet; and gorilla, in terms of the anatomy of its face. Of course, this is exactly what we should expect of a fossil ape that lived just after the orangutan and Africa ape lineages branched off from one another: a little bit of what would develop in each living species. So how do we know that *Dryopithecus* is related to the African apes and humans, a crucial part of my theory that the African apes and human lineage originated in Europe? The devil is in the details and there is no way to explain this without resorting to a little comparative anatomy.

As we saw in chapter 6, orangutans and *Sivapithecus* share a completely unique anatomy of the face that convinces nearly all researchers that *Sivapithecus* is, if not the direct ancestor, then at least an early member of the same subfamily as orangutans, the pongines. *Dryopithecus* has a more primitive face that is less distinctive, but it most closely resembles the face of modern gorillas, reduced to about one-quarter the size. Among the living great apes, gorillas have the shortest faces below the nose. This is hard to see in a living gorilla, which certainly appears to have a large, projecting face (see figure 1.6).

Cebus monkey

Chlorocebus monkey

PLATE 1. Old and New World monkeys. A New World monkey (*Cebus*, the capuchin monkey, *left*) and an Old World monkey (*Chlorocebus*, the vervet monkey, *right*). (Left image courtesy of Roger Sargent Wildlife Photography. Right image courtesy of Travis Steffens.)

PLATE 2. The apes. *Top left*, *Hylobates* (a gibbon); *top right*, *Pongo* (an orang-utan); *bottom left*, *Gorilla* (a gorilla); *bottom right*, *Pan* (a chimpanzee). (Top left image © Programme HURO from Wikimedia Commons. Top right image courtesy of Aukland Zoo. Bottom left image courtesy of Ken Ilio. Bottom right image © James Mollison.)

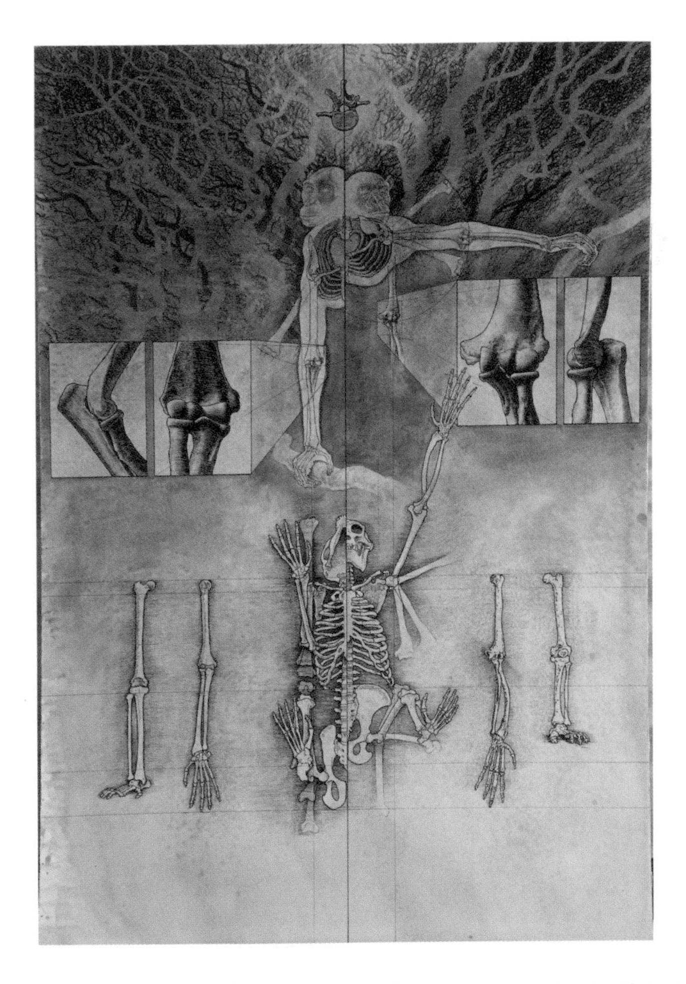

PLATE 3. John Gurche's illustration "Going Great Ape." A half-skeleton of an ape, specifically a chimpanzee (*left*) and that of an Old World monkey (*right*). Inserts show the structure of the elbow joints and the lengths of the arms and legs. Note the differences in limb position, chest shape, and lower-back length. (Image © John Anthony Gurche.)

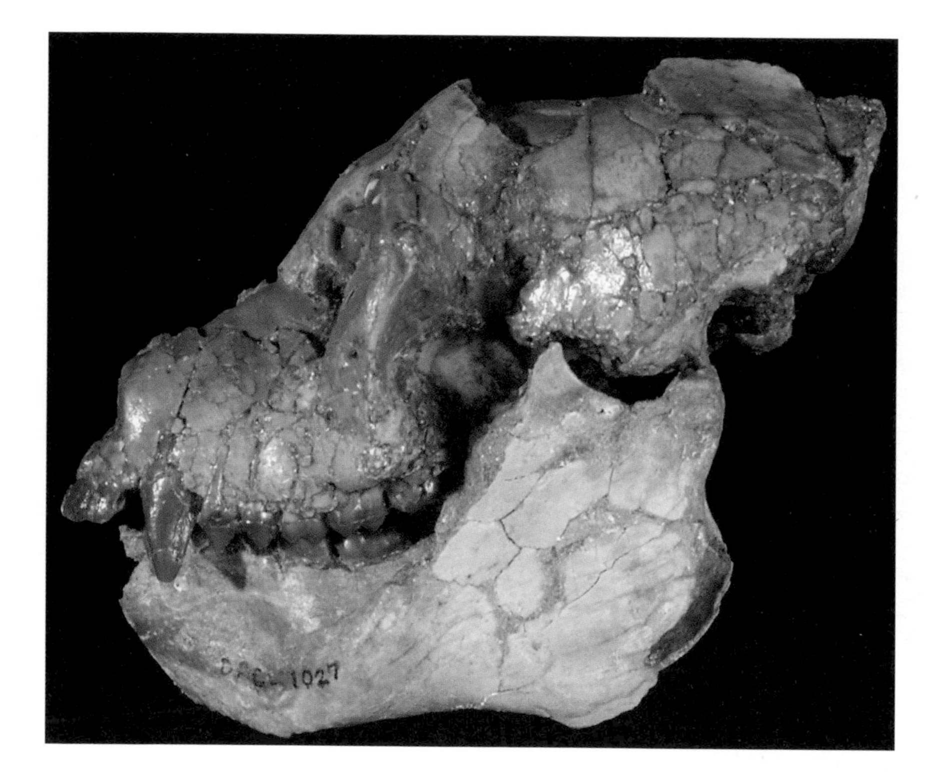

PLATE 4. Skull of a male *Aegyptopithecus*. The cranium and mandible are from different individuals. Note the small brain case and long snout. (Image © 2007 National Academy of Sciences, U.S.A., from Simons et al. 2007. "A remarkable female cranium of the early Oligocene anthropoid *Aegyptopithecus zeuxis* [Catarrhini, Propliopithecidae]." *PNAS* 104[21]: 8731–8736.)

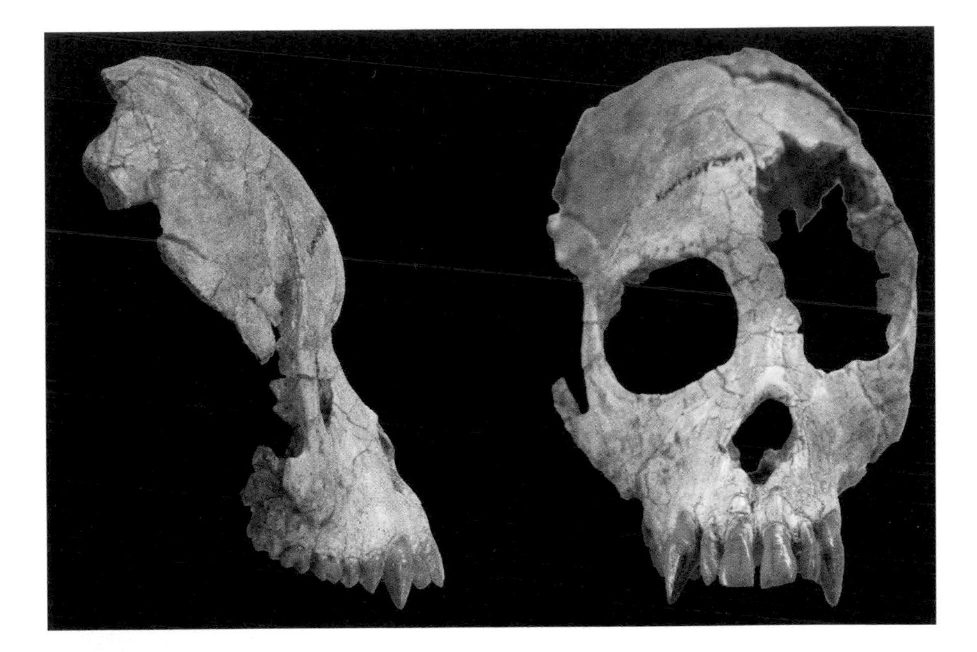

PLATE 5. The cranium of *Ekembo*. This is one of the best preserved skulls from the entire Miocene, a female that belongs to the species *Ekembo nyanzae*, from Rusinga Island, Kenya. Note the larger brain and smaller snout compared with *Aegyptopithecus*. (Images by author.)

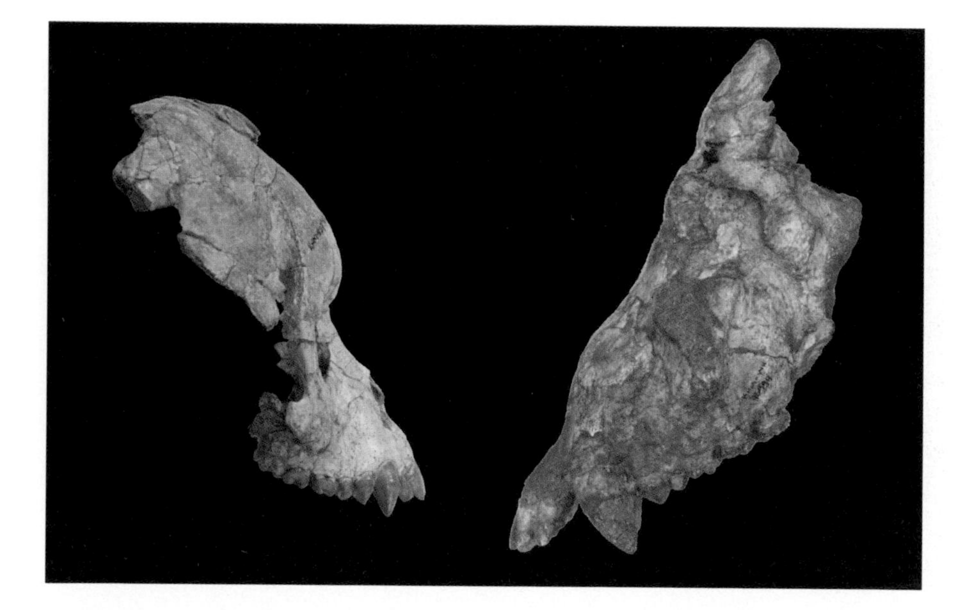

PLATE 6. A comparison of the crania of *Ekembo* and *Afropithecus*. *Afropithecus* (*left*) appears to have a long, strongly inclined snout but much of this is a result of distortion after the fossil was deposited. This is a common problem in paleontology. However, the differences in the development of the jaws and their associated muscles, which are much stronger in *Afropithecus*, are real. (Images by author.)

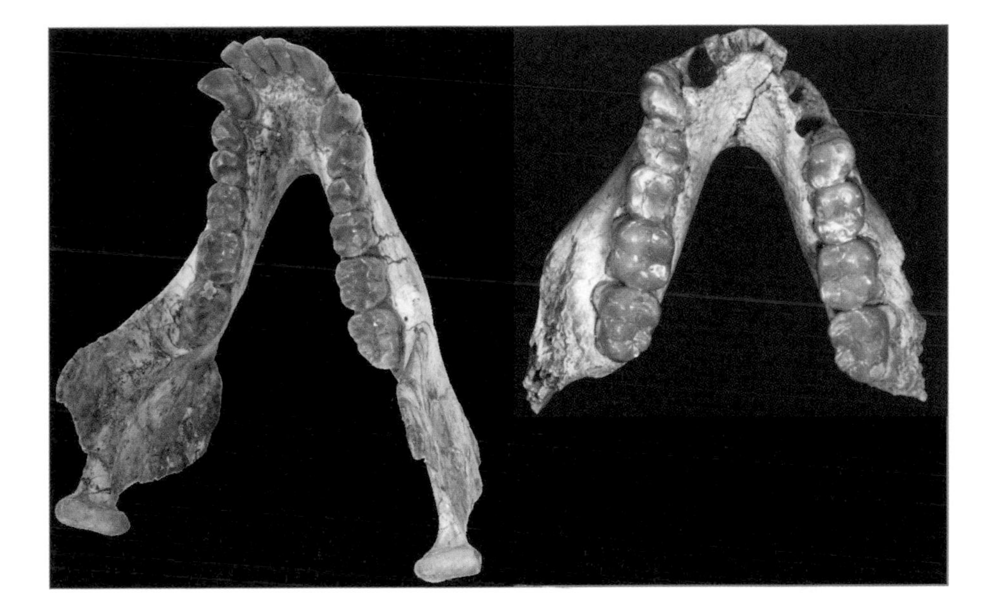

PLATE 7. Comparison of lower jaws of *Ekembo* (*left*) and Griphopithecus (*right*). Note the more massive construction of the mandible and the larger back teeth in *Griphopithecus*. Both individuals are female, which we can determine based on their canine morphology. Note also the reduction in the development of the buccal cingula in *Griphopithecus*. (Images by author.)

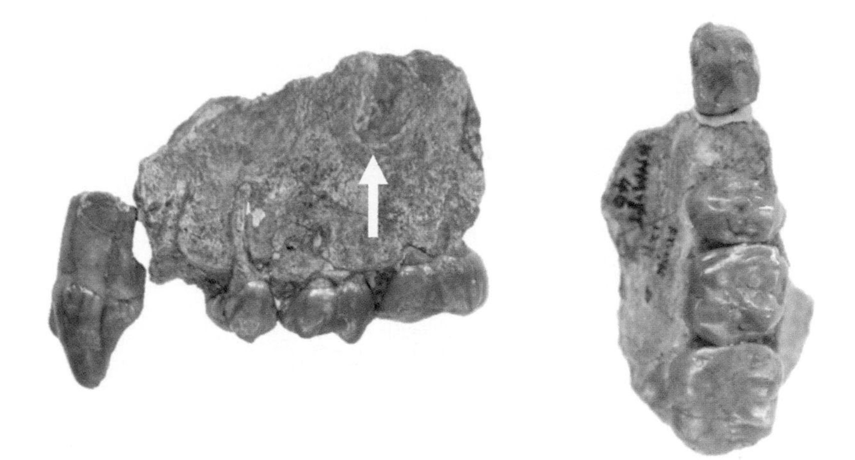

PLATE 8. Maxilla of *Kenyapithecus* with the canine. Note the small canine of typical female morphology and the somewhat higher placement of the zygomatic root (*arrow*). *Left*, left side view; *right*, view of the tooth surfaces with the canine and P^4 to M^2. (Images by author.)

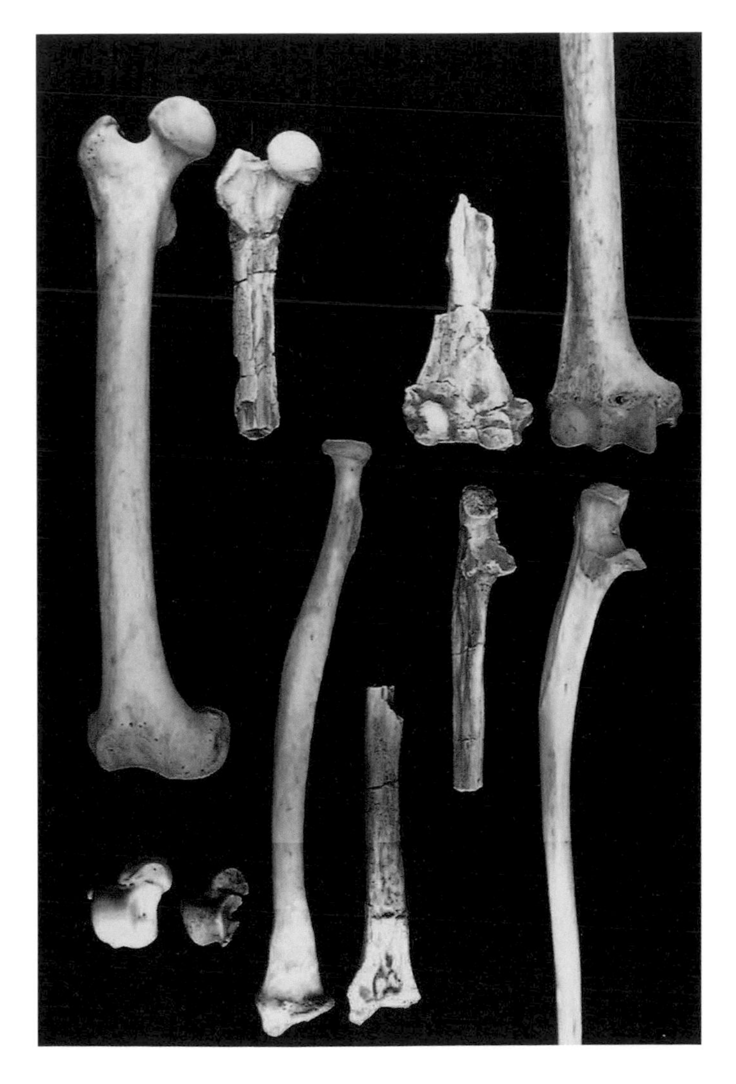

PLATE 9. Comparisons of the fore and hindlimbs of *Nacholapithecus* and a chimpanzee. The *Nacholapithecus* specimens (casts) are the more fragmentary ones in this image. (Images courtesy of Masato Nakatsukasa.)

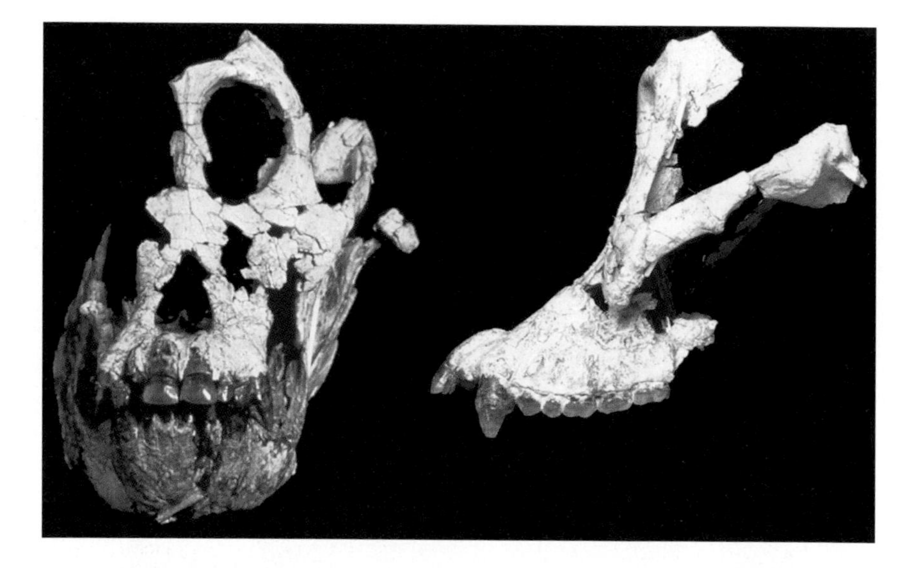

PLATE 10. Face of *Sivapithecus*. *Sivapithecus* in front and side views. Note the differences in facial contours and length. (Images courtesy of Milford Wolpoff.)

PLATE 11. The Rudabánya excavation. This is the R.II site, where most of the fossils from Rudabánya were discovered. (Image by author.)

PLATE 12. Skull of Gabi. *Left*, front (three-quarter) view; *right*, side view. (Images by author.)

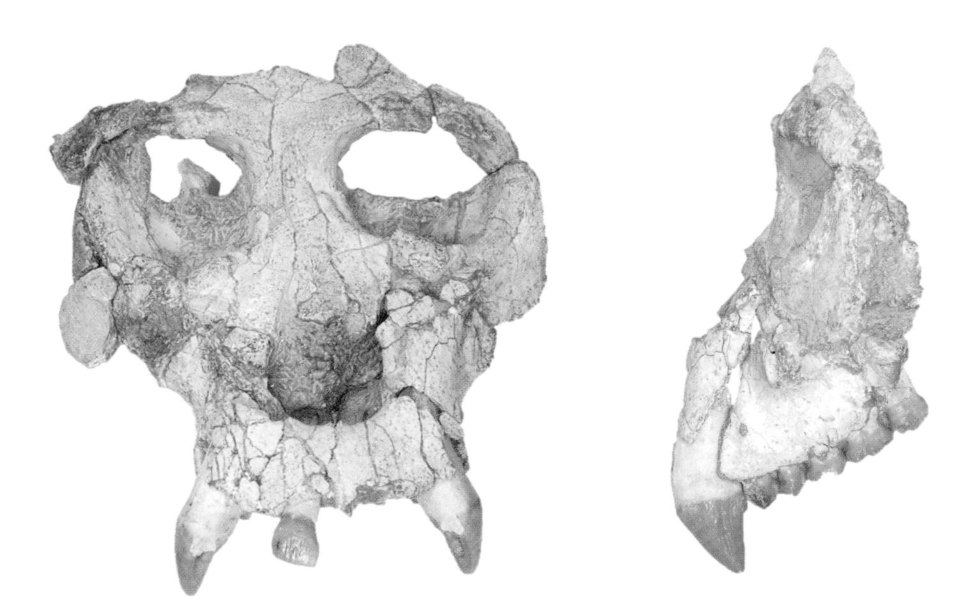

PLATE 13. Face of *Pierolapithecus*. *Left*, frontal view; *right*, side view. (Images by author.)

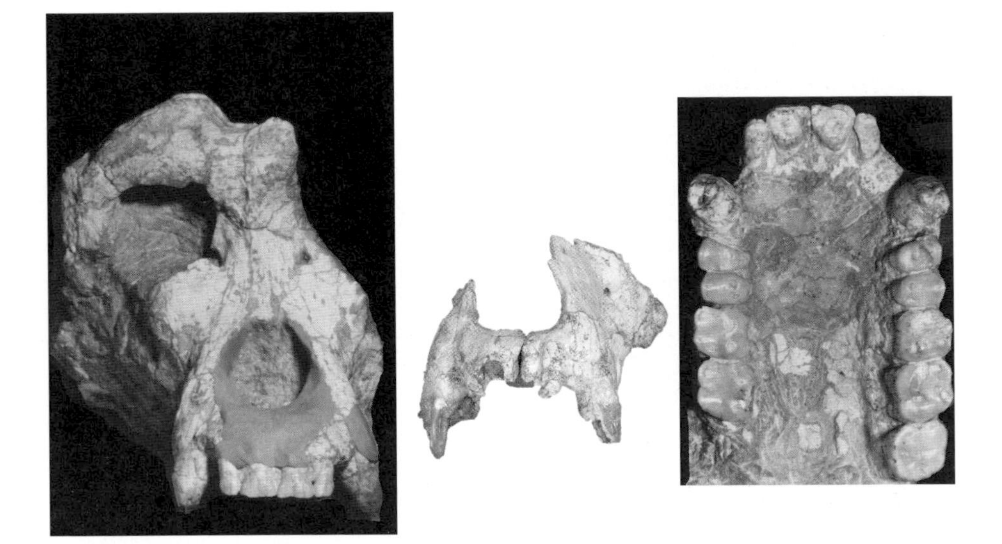

PLATE 14. *Ouranopithecus* and *Dryopithecus*. *Left*, frontal view of *Ouranopithecus*; *middle*, frontal view of *Dryopithecus* (Valles Penedes); *right*, occlusal view of *Ouranopithecus*. I have filled in some of the lower parts of the palate and premaxilla of *Ouranopithecus* with clay to try and reconstruct the shape of the nasal aperture. Like *Dryopithecus* it appears quite gorilla-like, with a broad, flat base and a large aperture. The view of the palate shows the large, flat molars and the relatively small canines of *Ouranopithecus*. (Images by author.)

PLATE 15. *Oreopithecus. Left*, the skeleton (cast) in the slab of coal in which it was found. Note the long limbs, although the length of the forelimb is exaggerated because one part, the radius and hand, has moved up. It should be next to the ulna, which is the bone that forms the sharp angle of the elbow (*arrow*). *Upper right*, the original specimen laid out. *Lower left*, a female palate (cast). Note the high cusps and sharp crests. *Lower right*, a close-up of an *Oreopithecus* upper molar. (Left image courtesy of Philippe Lopes. Upper right image © Ghedoghedo from Wikimedia Commons. Lower left image and lower right image by author.)

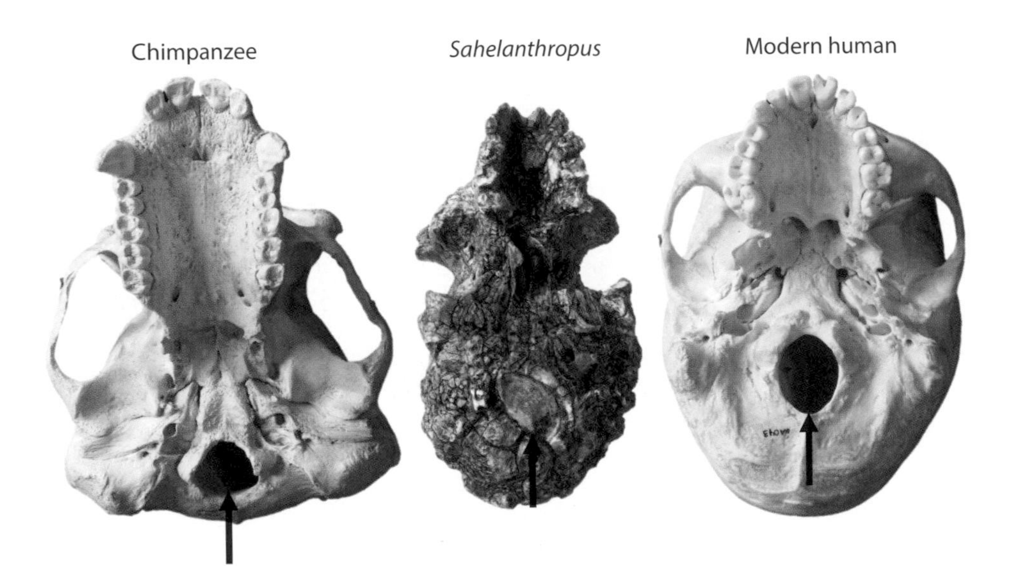

PLATE 16. Basicrania from a chimpanzee, *Sahelanthropus*, and a human. Note the more forward position of the foramen magnum in *Sahelanthropus* and humans (*arrows*). (Images by author.)

Much of this shape is caused by the larger canine teeth, which have roots that are so big they need a large face just to house them. Dentists sometimes refer to the canines as eye teeth because the roots run up the face close to the bottom of the eyes, even in humans. In gorillas they are much larger and have a dramatic effect on the shape of the face. The forward projection of the gorilla face almost stops below the nose because the front part of the upper jaws that houses the front or incisor teeth is short and vertically oriented. In chimpanzees and our ancestors the australopithecines, this part of the face is more projecting, probably because the incisor teeth are larger compared to body size. Chimps have larger incisors because they tend to prepare foods more intensively with these teeth, for example, peeling tough husks off fruit. Gorillas rely more heavily on their molars to slice through the fibrous foods that they eat.

The premaxilla is the part of the upper jaw or palate that holds the incisors. In gorillas the premaxilla is short and upright (see figure 1.6); the configuration in dryopithecins is very similar and differs from what we see in orangutans and chimpanzees. I interpret these similarities in morphology to mean that gorillas and dryopithecins share attributes of the face not found in orangutans, which have very specialized faces, or in chimpanzees and australopithecines, which have evolved a more elongated premaxilla. So among the African apes and humans, gorillas are the most primitive in terms of the anatomy of their palates and they are very similar to dryopithecins.

The teeth of *Dryopithecus* are similar to those of African apes. The enamel covering of the teeth, which was discussed earlier, is thin, as it is in African apes, and the teeth have chewing surfaces that very closely resemble those of African apes, especially chimpanzees, which have broad crushing basins and prominent rounded cusps, as opposed to the flatter, more thickly enameled teeth of orangs.

In other parts of the face there are also clues to the relations of dryopithecins. As we will see when we look at more recent species, the upper part of the face around the eyes is uniquely like those of African apes, as is the area where the lower jaw (mandible) connects to the upper braincase. The limb bones of *Dryopithecus* are also more like those of African apes than orangutans. The humerus that Lartet described is nearly identical to that of a chimpanzee, and a more complete skeleton from Spain described in 2004 also bears witness to a connection with African apes. This specimen, which its discoverers call *Pierolapithecus* 1, is from a place called Hostalets de Pierola, in the Vallés Penédés. The teeth of this species are nearly identical to those of *Dryopithecus* from elsewhere in Spain and France, although the face looks different. The eyes are tilted strongly upward, a similarity the discoverers see with *Afropithecus*, which we met in chapter 3. In my view however, this is an artificial similarity caused by distortion in the Spanish fossil. In all other respects it looks very similar to *Dryopithecus* (plate 13).

The Hostalets specimen is very important because it consists of an associated partial skeleton. As noted in chapter 2, nearly all fossil ape species are identified initially by their jaws and teeth. Jaws, especially mandibles and teeth, are more durable than most other bones of the body. Mandibles must withstand the very high forces generated by chewing, so they have to be strong. Teeth are covered with enamel, the hardest substance the body produces, and so they preserve well in the fossil record. Unless limb bones are found in direct association with jaws and teeth, it is very difficult to know to which ape they belong (remember the story of *Paidopithex* in chapter 2).

There are several aspects of the body of *Dryopithecus* that are very different from both pliopithecoids and living monkeys. The limb bones are poorly preserved, so we cannot directly measure their lengths to say if the arms were longer than the legs, as is the case in all apes (monkeys have arms and legs of roughly equal length). But the lower back is represented by a nicely preserved vertebra, and the wrist also is well preserved. One could not ask for better areas of the body to see look for differences between monkeys and apes.

As I noted in chapter 1, the backbones of monkeys and apes are very different. Monkeys have seven bones in their lower back, between their rib cages and their hips, which are called the lumbar vertebrae. Great apes have between three and four (more often three), gibbons and siamangs have five, and humans also have five. Why the differences? To review, monkeys have long backs that give them a great deal of flexibility, agility, and speed in running along the tops of branches and on the ground. The long lower back separates the front and hind limbs, which increases their stride length in the same way that persons with long legs take fewer steps and can swing their legs at lower frequencies and still cover the same distance at the same speed as persons with shorter legs, who have to take more steps. In rapid locomotion a monkey can arch its back downward, further separating its front and back limbs, and they can also more easily twist their bodies as they leap along the branches. Apes have short lower backs. Because they are short and have fewer bones, ape lower backs are very stiff and solid. In gibbons and siamangs, the loss of two of the lumbar vertebrae may be related to their more upright posture. Gibbons spend much of their time hanging below branches and swinging from them, and their backbones are more vertically oriented. They do not need to have widely separated upper and lower limbs, and the reduced number of lumbar vertebrae makes their backs more stable. Great apes experience a further reduction in the number of lumbar vertebrae, probably because apes are much heavier than monkeys, gibbons, and siamangs and need the greater stability. All great apes most often locomote using all four limbs, by grasping onto multiple branches with their hands and feet, so they need a stiff, solid connection between the two. It is not clear why humans appear to have traded a thoracic vertebrae for a lumbar (apes have 13 thoracic, or rib cage vertebrae, whereas humans have 12). Humans may need a longer lower back to facilitate hip rotation, which is necessary for a smooth, striding bipedal gait. This helps us to swing our feet in front of our bodies rather than off to the side.

What does the Hostalets specimen tell us about the early evolution of the great-ape back? Unfortunately we cannot say how many vertebrae that specimen of *Dryopithecus* possessed, but the one well-preserved lumbar vertebra is apelike in structure. Once again, to remind you about vertebrate comparative anatomy, most mammalian vertebrae have bony projections that stick out to the sides and the back, which are for the attachment of the muscles that move and support the backbone. The ones that stick out to the side, the transverse processes, are positioned more toward the front of the vertebrae in monkeys and more toward the back in apes. This reconfiguration of the transverse processes in apes, which was discussed when we were looking at *Morotopithecus*, is thought to increase the stiffness of the lower back by repositioning some of the muscles. In the lumbar vertebra of *Dryopithecus* from Hostalets de Pierola, the transverse processes are more apelike than even in the Moroto specimen we met earlier. I interpret this to mean that *Dryopithecus* had a lower back more like that of living great apes, meaning that it was most probably more orthograde (having a relatively vertically oriented backbone). This is the earliest evidence of a great ape's lower back, and the fact that it is found in Europe is very interesting, since no great apes live in Europe today and the ancestral homeland of the great apes and humans has always been assumed to be in Africa (chapter 6).

The other part of this skeleton that is very interesting is the wrist. There are eight bones in the wrists of African apes and humans and nine in orangutans, gibbons, and most other primates. This part of the body in primates is among the most primitive. What I mean by that is that primates have either eight or nine wristbones because we resemble the earliest mammals, which also had a large number of wristbones. Many living mammals have fewer wristbones, which makes their wrists more stable in rapid locomotion but with limited flexibility. One of the hallmarks of the primates is the retention of a primitive limb structure that enhances flexibility, enabling primates to grasp at branches from various angles and orientations while at the same time allowing them to run rapidly. Nevertheless, there are differences among the primates. Monkeys have more stable wrists than apes, because both bones of the forearm connect with bones on each side of the wrist.[3] This structure makes the monkey wrist more stable by forming the equivalent of a carpenter's mortise joint, the kind of joint that holds well-made drawers together. Monkeys can move their wrists from side to side to some extent, as we do, but they are mainly designed for flexion and extension, or bending forward and backward (chapter 2). Quadrupedal animals generally have this type of wrist joint so that they can move rapidly without their wrists deviating to one side or the other. In gibbons and siamangs there is a partial connection to the pinky side of the wrist, and in great apes there is no bony contact at all. This allows apes and humans to rotate our wrists much more extensively than any other mammal, which you can see on your own wrist by holding your palm toward your face and bending your wrist toward the pinky side. That is ulnar deviation, and the degree to which it is developed is unique to great apes and humans. It allows us to place our wrists in a greater number of different positions, which is important when you are climbing trees and especially when you are suspending your body weight below the branches. The increased mobility helps prevent damage to the wrist when the body twists around an arm that is supporting it. The *Dryopithecus* specimen from Hostalets preserves one of the bones from the ulnar or pinky side of the wrist, the triquetrum, and it lacks the facet or joint surface for the ulna, so it is just like the wrists of great apes in that regard.

The other bones of the skeleton of *Dryopithecus* closely resemble those of living great apes more generally and tell us that this ape led a similar life style. I mentioned that the humerus of *Dryopithecus fontani* is long, slender, and curved, just like that of a chimpanzee. The Hostalets specimen preserves finger bones that are a bit short compared with those of living great apes and later dryopithecins but still show unambiguous signs that this animal moved around in the trees mainly by suspension, even if not to the same extent as in living apes.

The jaws and teeth of *Dryopithecus* from the nearly 12-million-year-old deposits of northern Spain and France have been extensively studied for clues concerning the diet of these early hominines. As I mentioned earlier, researchers can deduce much about what extinct animals ate from the traces left on their teeth and the structure of the teeth themselves. In the case of *Dryopithecus*, the molars or grinding teeth have prominent cusps that surround shallow, broad basins, and the enamel that covers the teeth is thin. This is precisely the kind of tooth that living chimpanzees have, and their teeth have evolved to crush and grind relatively soft foods, such as ripe fruits. Molars have ridges that connect the cusps, and these are also similar to what we see in living chimpanzees. Gorillas, for example, have much taller cusps and more strongly developed ridges that they use to chop and slice the more fibrous vegetation that forms the bulk of their diet. Orangutans, in contrast, have low-cusped teeth with very broad basins and thicker enamel for more powerful crushing of harder and tougher food items, such as fruits that are protected

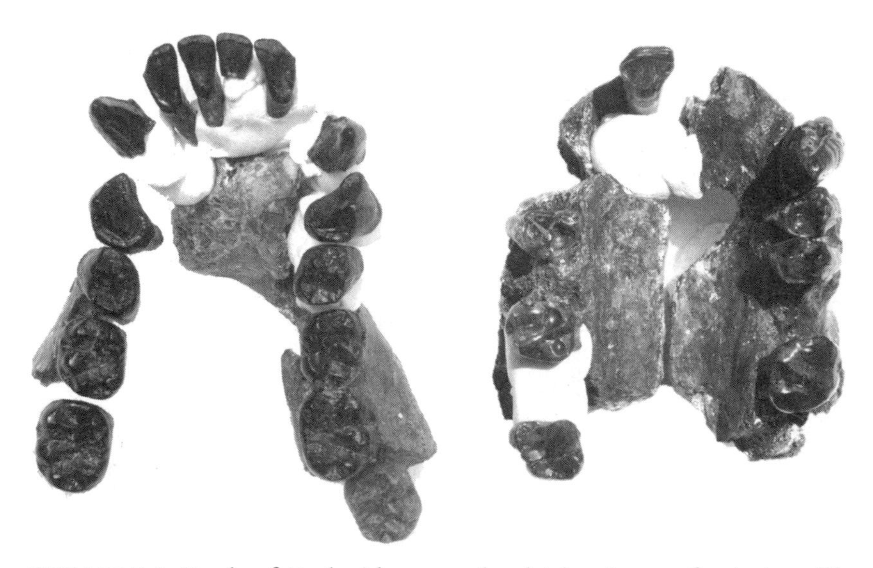

FIGURE 7.1. Teeth of *Rudapithecus*, occlusal (chewing surface) view. The mandible (*left*) is from a younger individual. Note the differences in degree of wear and the unerupted last molar. (Images by author.)

by husks. So we can conclude that the jaws and teeth of *Dryopithecus* are fairly generalized, lacking the specialized developments of orangutan and gorilla dentition. Its diet was probably closer to that of a chimpanzee (figure 7.1).

We know little else about *Dryopithecus*, the oldest and most primitive of the subfamily we share with chimpanzees and gorillas. Other great apes from Europe, however, give us a more complete understanding of this phase of our evolutionary history. To end this chapter I want to say a little more about how paleontologists know about the diet of extinct organisms. Paleoanthropologists and paleontologists generally and really do appreciate the saying "you are what you eat." We talked earlier about the structure of teeth. The teeth of foliage-eating animals, including primates, have tall pointy cusps and sharp crests. Fruit-eating primates have low rounded cusps and broad basins. They also tend to have larger front teeth (incisors), which they use as tools to remove the outer coverings of many fruit. Bill Hylander is a pioneer in this field. Thirty years ago he revealed the relationship between incisor morphology and diet in primates, and he trained many students, now leaders in the field, in primate jaw biomechanics.

In addition to this arsenal of analogies to living primates of known diet, paleoanthropologists occasionally recover direct evidence of diet on teeth, as in the case of the phytoliths mentioned in chapter 6. More commonly, paleoanthropologists examine the wear patterns on teeth. We have noticed that in living animals those that eat fruits with hard or tough coverings tend to have more pits—short, variably deep defects caused by the food they eat. Those that eat fibrous foods tend to have more scratches—long and usually shallow marks. Both pits and scratches can be counted and compared by making replicas of the teeth and examining them under a scanning electron microscope. More recently, paleoanthropologists have used software originally developed to quantify landscapes to characterize the "hills" and "valleys" of teeth and how they differ between frugivores and folivores. Finally, paleoanthropologists use techniques borrowed from geochemistry to analyze the chemical composition of teeth and bones. Different diets leave different isotopic signatures (different ratios of various forms of elements such as carbon, oxygen, strontium, and calcium). Using all these techniques, paleoanthropologists have been able to describe the diets of many fossil species, from apes to Neandertals.

CHAPTER 8
THE DESCENDANTS OF *DRYOPITHECUS*

In Europe, *Dryopithecus* appears to have evolved into several new genera that were more like living great apes but were also unique to their geographic and ecological circumstances. About 2 million years after the first occurrence of *Dryopithecus*, we find *Hispanopithecus* in Spain and *Rudapithecus* in Hungary. Half a million years later, we have *Ouranopithecus* in Greece. We are fortunate that for two of these three early great apes we also have partial skeletons.

HISPANOPITHECUS

The skeleton of *Hispanopithecus*, the discovery of which I described at the beginning of chapter 7, is very important because it includes well-preserved cranial remains and portions of the upper and lower limbs and the backbone. Remember that the limb bones of most fossil organisms can usually only be assigned to a species if they are associated with jaws and teeth or other diagnostic skull remains.

My association with *Hispanopithecus* started in 1983 when I visited Professor Miquel Crusafont-Pairó in Sabadell, Spain. Crusafont-Pairó was a revered vertebrate paleontologist in Catalonia, the personification of the local tradition of natural history and a highly respected figure in the field in general. When I met him he was retired and dying of cancer, but he was incredibly gracious, coming to greet me and show me around. I will always remember his slow walk, cane in hand, down the corridors of what is now the Institut Paleontologia Miquel Crusafont-Pairó, impeccably dressed in the local style of a gentleman from the mid-twentieth century, a class act to the end. I spent a very pleasant afternoon with Professor Crusafont, and he permitted me to photograph and make molds of all the ape fossils in his care. Sadly this sort of access, especially for young researchers, is increasingly rare. I was and continue to be moved and grateful for his kindness and openness and the degree to which he was genuinely interested in my research. When the time came to name a new species of *Dryopithecus* (now *Hispanopithecus*) based on my analysis of the specimens from Spain, I knew immediately that it would have to be *Hispanopithecus crusafonti*.

With *Hispanopithecus* we have a face, consisting of the upper jaw, or maxilla, and fragments of the midface connecting the upper jaw to the area around the eye sockets. The specimen also includes part of the brow ridges above the eyes, but none of the braincase and no mandible. Along with these cranial remains, we also have portions of the upper and lower limbs and a part of the backbone. The finger bones are very long and curved, as in living orangutans, although the bones of the palm (the metacarpals) are shorter than we would expect for an animal with such long fingers (figure 8.1).

You can tell a lot about the way an animal moves about by the anatomy of its extremities. A closer look at the finger bones of *Hispanopithecus* reveals telltale signs of habit. Long, curved phalanges, or finger and toe bones, are found only in primates that are suspensory and spend most of their time in the trees. In addition, these bones have ridges on the sides of their shafts that mark the attachment of sheaths that form little tunnels to guide the powerful flexors, the muscles that bend the fingers as you would do when making a fist (chapter 2). However, the metacarpals are also usually elongated in suspensory primates, but not so much in *Hispanopithecus*, which simply tells us that it has its own unique anatomy. The behavioral significance of this difference is not clear. Other anatomical attributes are also consistent with suspension. The arms are also long compared with the legs, a feature unique to the apes

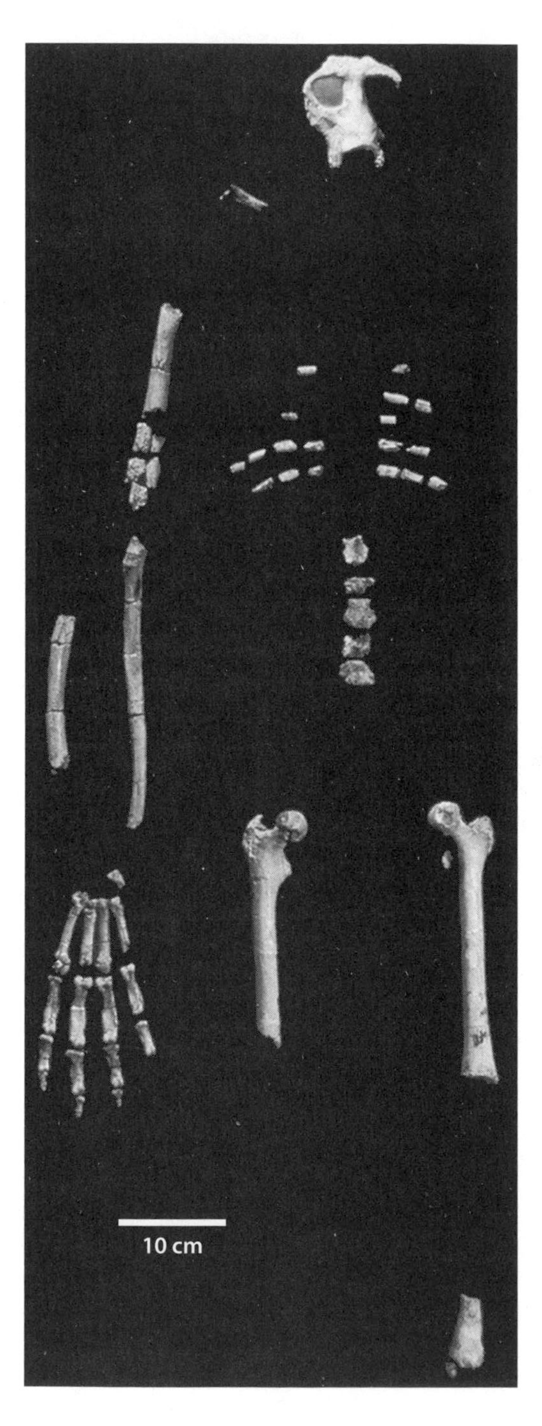

FIGURE 8.1. The skeleton of *Hispanopithecus*. The face, the forearm, the thigh bone (femur), and the hand are especially well preserved. The face and the end of the tibia (part of the ankle joint) were found in 1992 while I was codirector of the project. The rest of the skeleton was found by Salvador Moyà-Solà and his group the following year. (Image courtesy of David M. Alba and Salvador Moyà-Solà.)

among primates. Most important, a lumbar vertebra is preserved, and it looks like a typical great ape lumbar vertebra, with the transverse processes positioned toward the back.

The *Hispanopithecus* skeleton is one of those specimens I mentioned earlier that has a femur that looks quite different from that of *Paidopithecus* from Eppelsheim. It is identical in all important details to the femora of living great apes. Great apes in general have short, stout femora with large femoral heads. The head of the femur fits into a large socket in the hip bone, or pelvis, called the acetabulum, forming a very stable ball-and-socket joint. In great apes the hip joints are large, and the ball of the hip joint forms a nearly perfect sphere. Also in great apes, the muscles that surround the hip are also placed such that mobility at the hip joint is maximized. The arrangement in the hip is similar to what we saw in the wrist. Great apes have more flexibility and are able to place their legs in a great diversity of positions, which is very helpful in climbing and spreading their body weight across a number of supports.

Orangutans are the champions of that skill. They can support their bodies on two branches with both legs sticking out at 90-degree angles to their trunks. In addition to having nearly perfect spheres for femoral heads, orangutans also lack a ligament on the surface of the femoral head that all other primates and the vast majority of other mammals have: the ligamentum capitis, which attaches to the fovea capitis on the surface of the femoral head. The absence of this ligament gives orangutans even more mobility at the hip joint. What a specialization that is, and perhaps it is an indication that orangs evolved their extreme form of suspensory behavior separately.

In *Hispanopithecus* the femoral head forms a large sphere, and it is separated from the rest of the bone by a stout neck. Attachment sites for the muscles that move the hip are placed lower on the shaft to allow for more movement at the hip joint, as is the case in living great apes. Unlike orangutans however, *Hispanopithecus* and all known fossil apes have a fovea capitis. All this evidence shows without any doubt that *Hispanopithecus* moved about in its environment in a manner very similar to the way great apes move today. It was an excellent climber and could easily suspend its body from branches as it swung across the canopy.

In *Hispanopithecus* we also see the development of an African ape characteristic that is not known for *Dryopithecus* (because it is not preserved, though I think it was probably there). Most mammal skulls have cavities called sinuses that are filled with air, and they emanate from the area between the eye sockets: the ethmoid region. In some mammals these air sinuses, called paranasal sinuses, are small and don't extend far, and in others they can be extremely elaborate. It turns out that in hominoids there are three basic patterns to the development of the paranasal sinuses. Hylobatids have relatively small sinuses that extend from the ethmoid to an area behind the eyes in a bone called the sphenoid. From the sphenoid there is a secondary extension of the sinus system into the frontal bone, which is often hollowed out above the level of the eye sockets. This configuration is also found in a number of monkeys. In orangutans the sinus system is even more limited and does not extend as far as the frontal bone. In African apes and humans, the sinus system is extensive, having large and direct connections between the ethmoid and frontal portions. *Hispanopithecus* has a well-preserved ethmoidal region with a large sinus that is continuous with the frontal sinus, a clear similarity with African apes. Once again, this is a bit of a surprise, since the African apes were long assumed to have been confined for their entire evolutionary history to Africa. This is evidence of a more complex evolutionary history (figure 8.2).

The frontal sinuses of *Hispanopithecus* are small compared with those of African apes and humans, which is to be expected in an early member of our subfamily. Another feature of the skull of *Hispanopithecus* that is like those of African apes is the brow ridges, which are also smaller, or less well developed than in living forms, but definitely visible. The brow ridge is represented by low, long mounds of bone just above the eyes in the fossil ape. Orangutans have no brow ridges, whereas in other living apes and many fossil humans the ridges are strongly developed. *Hispanopithecus* and the other descendent of *Dryopithecus*, *Rudapithecus*, both show the early development of this feature of the skull.

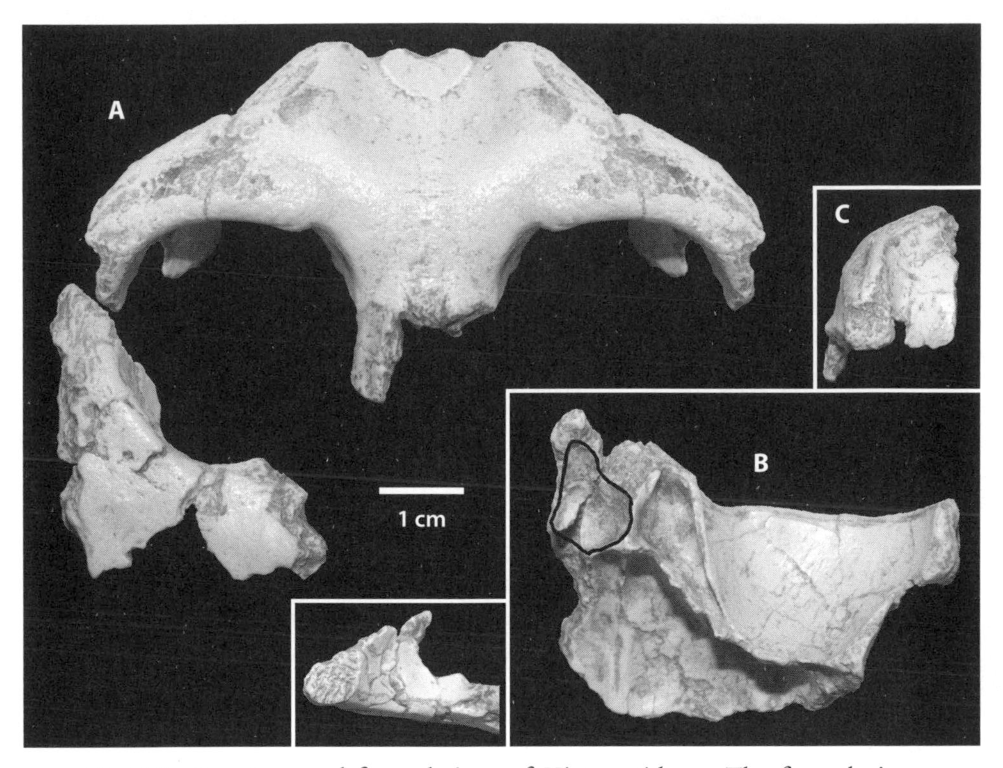

FIGURE 8.2. Face and frontal sinus of *Hispanopithecus*. The frontal view (*A*) is mirror imaged from the left side to make a complete frontal bone. Note the wide space between the orbits in view A. View B is the same bone from behind, revealing an extensive frontal sinus (outlined on one side). View C is a side view of the frontal bone showing the incipient development of the brow ridges. (Images by author.)

Piecing together the story of early hominine evolution involves a large number of small fragments of evidence that have to be integrated. The interpretation I am describing is not universally accepted, but it is the most logical and consistent explanation that I can come up with, and it is consistent with all the facts. However, until we get a confession (for example, via DNA evidence) out of one of these fossil apes, we will never be sure of what actually happened. We do not yet have the technology to extract DNA from these fossils, but that may be possible one day.

RUDAPITHECUS

On the other side of Europe we have equally impressive fossils of the close relative of *Hispanopithecus*, *Rudapithecus*. The history of the discovery of the fossils from the *Rudapithecus* site, Rudabánya, is interesting. The Ruda is a stream running through the hilly country of northeastern Hungary, and the town of Rudabánya exists as a result of the ore mine that was once the foundation of the local economy (*bánya* means "mine" and is a common suffix for towns associated with mines in Hungary). Human occupation of the area is very ancient, with evidence of mining going back thousands of years. Until the end of the 1970s, Rudabánya was relatively affluent by Cold War–era, Soviet-bloc standards, because of the importance of the mine as a source of iron, copper, and other resources for the Soviet economy. Sometime between 1965 and 1967, during the heyday of the mine, a fossil site was discovered by accident while an open pit was being expanded. The geologists who run mining operations are interested in the rocks that contain the ores and not the rocks and dirt that lie above them, even if they are rich in fossils. So it was very lucky that one of these geologists, Gábor Hernyák, whom we met earlier, had a strong amateur interest in fossils, and he intervened when the mine activities exposed fossil-bearing rocks. One day in the mid 1960s, Hernyák found a piece of a mandible that he knew was interesting, although he was not sure why. Rudabánya was famous for the fossils that could be found on the periphery of the mine, and Hernyák knew that the most prominent Hungarian paleontologist of the day, Professor Miklós Kretzoi, was interested in learning about anything new that came from Rudabánya. So Hernyák delivered the jaw to Kretzoi, and the rest is history.

Kretzoi immediately recognized the specimen as an ape jaw. At the time, the prevailing theory of ape and human evolution was that the human lineage branched off from the great ape lineage as early as 15 to 20 million years ago and that the earliest members of the human lineage were represented by the *Ramapithecus*. Remember the *Ramapithecus* saga (chapter 5)? *Ramapithecus* was thought to be ancestral to humans. As noted, a major reason for this interpretation was the apparent presence of a small canine, a key human feature. Kretzoi saw in the small jaw from Rudabánya evidence of this reduction in canine size, and he attributed the specimen to a new genus closely related to *Ramapithecus*, *Rudapithecus*. Kretzoi believed *Rudapithecus* to represent the first evidence of this ancient branch of the human family in Europe.

As we now know, the real story is much more complicated. As described in chapter 5, *Ramapithecus* is now thought to be composed of female specimens more properly attributed to *Sivapithecus*. The small canines reflect the fact that all the specimens are female and not that they are more humanlike.

As more researchers examined the specimens, it became clear that *Rudapithecus* lacked the powerfully built jaws and thickly enameled teeth of ramapithecines. In his enthusiasm to recognize an early human ancestor from Europe, Kretzoi had failed to see the much stronger similarities between *Rudapithecus* and *Dryopithecus*. We now know that *Rudapithecus* is a dryopithecin and not a sivapithecin, but the initial work by Kretzoi was certainly very important in bringing to the awareness of the scientific community the existence of a fossil ape from Europe with affinities to modern species from Africa.

I was very fond of Miklós Kretzoi. Most people who knew him called him Miklós Bácsi, roughly translated as Uncle Miklós, a term of affection and respect in Hungary. He let me stay at his apartment when I was a 21-year-old student living for the first time behind the Iron Curtain, and he and his wife Marika actually left me on my own in their place for a week to work on primate fossils from Rudabánya while they vacationed at the holiday mecca of Hungary, Lake Balaton. It was a touching and generous gesture to make to a kid they had never met before; it launched my career as a Miocene-ape investigator and triggered my soft spot for the older generation of vertebrate paleontologists.

In my dissertation I had lumped *Rudapithecus* with *Dryopithecus*, and in subsequent years, with the discovery of new and more complete fossils, I came to recognize the connection with African apes and humans. Most recently I have returned to the idea that *Rudapithecus* and *Dryopithecus* are distinct genera, with the former probably evolving from the latter. The paleontological sites surrounding the mine at Rudabánya have yielded the largest collection of well-preserved cranial remains of any European great ape. A face similar in structure and preservation to the *Hispanopithecus* face is known from the Rudabánya, as are two other skulls with parts of the braincase preserved, unlike the specimens from Spain. In my 1999 excavations, we uncovered a cranium with most of the face and the braincase attached (Gabi, as described earlier; see plate 12). In 2006, the next year we returned to work at Rudabánya, we found her mandible and parts of her hip bones. These discoveries of a small *Rudapithecus* female made me realize that I had made a mistake in my attribution of other specimens. The small femora that we had found in 1998, which I had erroneously attributed to *Anapithecus*, also belonged to Gabi (the hip joints fit together perfectly), so we had a partial skeleton.

Gabi is an extremely important specimen for a number of reasons. There are two partial braincases of *Rudapithecus* and both are complete enough to provide the basis for an estimate of brain size. These are the only specimens in Europe for which this is possible. Gabi, with her more complete braincase, confirmed our previous estimate that the brain of *Rudapithecus* (which at the time we were still calling *Dryopithecus*) is the same size as that of a chimpanzee of equivalent body mass. This is the earliest direct evidence of chimpanzee-sized brains in the hominoid fossil record. Chimps and other living great apes have brains that are roughly one-quarter to one-third the size of modern human brains, but they are more than twice the size of any monkey's brain. So great apes have small brains compared to humans but very large brains compared to all other primates. And they have the intelligence to go with their large brains. We have an estimate of the brain size of *Proconsul*, which I mentioned earlier, and an indirect estimate of the brain size of another European ape, *Oreopithecus*, to which I shall return at the end of this chapter. Both are in the size range of baboons, much smaller than the smallest great ape brain (figure 8.3).

Rudapithecus, with its very large brain, is clearly a great ape not only from an evolutionary point of view, but also in terms of its biology and behavior. Big brains are metabolically very expensive to

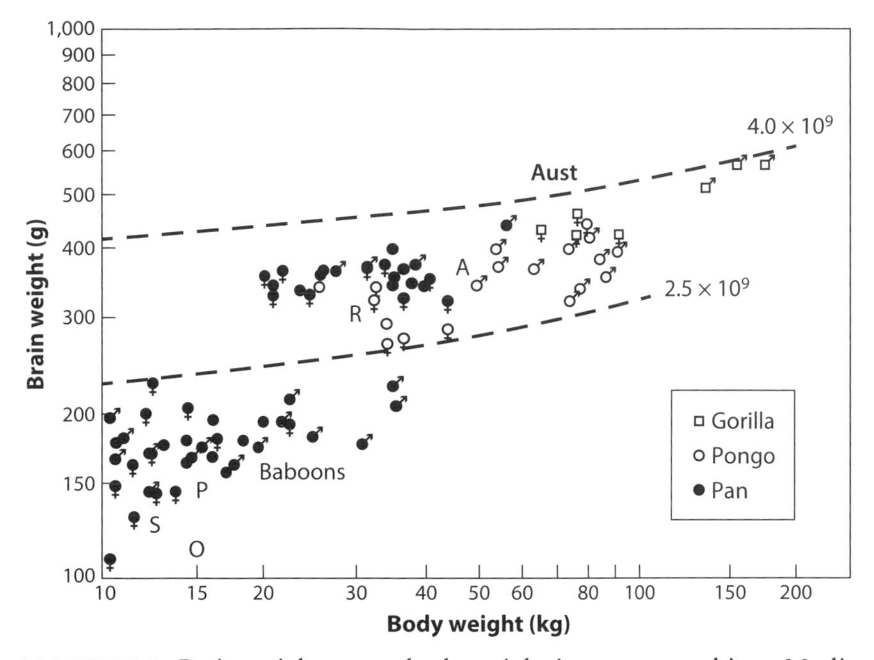

FIGURE 8.3. Brain weight versus body weight in some catarrhines. Modification of a graph published by Jerison in 1973 showing the position of australopithecines known at the time (*Aust*) relative to great apes and baboons. I have added *Rudapithecus* (*R*) and *Ardipthecus* (*A*), which fall well within the great ape cluster. Also added are *Ekembo* (*P*), *Symphalangus* (*S*), and *Oreopithecus* (*O*). *Rudapithecus* and *Ardipithecus* have achieved the same great-ape grade of encephalization, whereas *Ekembo* and siamangs are more like baboons, the most encephalized of the Old World monkeys. *Oreopithecus* is what I like to call a pinhead, a rare example in primate evolution of a lineage that has probably experienced a reduction in brain size.

maintain, and there have to be very compelling adaptive reasons for them to develop. In humans the brain consumes about 20% of our total intake of calories per day, although it accounts for just 2% of our total body mass. So the brain consumes, on average, ten times more than the average of the remaining tissues of our body. The payoff is intelligence. This fact alone makes it clear that the idea that humans use only a small portion of our brains is a myth. We simply can't afford to waste that amount of energy.

Many different mammal lineages have experienced brain-size increases up to a point, but none, except perhaps dolphins, to the extent seen in great apes and especially humans. The reason for this general trend is usefully explained in terms of a feedback loop, or more colorfully, an "arms race." There is a basic relationship between average brain size in a species and the level of cognitive complexity: species with larger brains tend to be more clever at problem solving, although the causal mechanism of this relationship is mysterious. We know, for example, that modern humans can differ in brain size by a two-to-one ratio; that is, a perfectly normal person can have a brain as big as 2000 cubic centimeters or as small as 1000 cubic centimeters, and there is no cognitive difference between the two (remember from chapter 1 Einstein's average brain size). However, when we compare species with species, animals that have brains twice the relative size of another animal's brain are generally more intelligent, after the effects of total body size are taken into account. Like an arms race, there is a tendency in brain evolution to spend more on brain mass, and presumably intelligence, in response to threats. Animals that have a greater capacity to predict what other animals will do, whether those other animals are predators or members of its social group, or to predict or understand and remember variations in their local environments, such as seasonal availability of resources, will be better able to survive. On the other hand, predators and/or other group members are also being challenged to keep pace with this increase in cognitive processing capacity, and their brains get bigger and they get smarter as well. This cycle will continue until a threshold is reached where too much energy is required to grow and maintain a healthy brain. Somehow primates have at least in part beaten that threshold, and the arms race among primates has yielded bigger brains than for any other group of mammals except odontocetes, that is, dolphins and other toothed whales.

Rudapithecus had achieved what we call a great-ape grade of brain evolution, meaning that it was like modern great apes in having a much larger brain relative to its body size than almost any other animal. The same was probably true of *Sivapithecus*, as noted in chapter 7, but based on indirect evidence probably *Hispanopithecus* and *Dryopithecus* as well. Because *Dryopithecus* and *Hispanopithecus* are more closely related to *Rudapithecus* than to *Sivapithecus*, if both *Rudapithecus* and *Sivapithecus* had great-ape-sized brains, we reason that the other dryopithecins likely did as well.

All of these fossil hominids probably benefitted from the presence of large brains in the same way as modern great apes do. As we saw in chapter 1, great apes generally have more complex social relationships and more involved strategies for finding food than most monkeys. They may have longer memories and a greater capacity to map their environments and its resources, which could be critical to tracking resources in over both the short and long term. They also have more elaborate systems of communication. Great ape babies are born at a relatively immature state, and they need more intensive parental care for longer periods than monkeys do. This necessity requires cooperation among adults to care for the infants, and it also provides the infant with learning opportunities while its brain is still growing, which has a direct effect on the way brain cells interconnect. Growing more of your brain after birth allows you to fine tune the structure of your brain to suit the circumstances of your social group and your ecological challenges. So, as with later hominids, the large brain of early hominids is almost certainly a sign that these animals were more cognitively complex than monkeys and possibly on a par with living great apes. We will return to the issue of brain size and adaptability when we have a look at the reasons the apes of Europe went extinct.

Big brains come with a price. In addition to the fact that they are metabolically very expensive, requiring a constant supply of high quality nourishment, there is another problem.

Humans are rare among mammals in that we suffer from a condition known as cephalopelvic disproportion. Simply put, women have difficulty giving birth, more so than in most mammals, because the baby's head is usually just barely smaller than the birth canal. In many cases the heads are actually too large to fit through the canal. Before modern medical interventions were developed, birthing was the leading cause of mortality in women of childbearing age. Caesarian sections (which have existed at least since the time of Julius Caesar, after whom they are named) are fairly routine (although tell that to a women who has just had one!) and are rarely followed by serious complications.

The reason that humans exhibit cephalopelvic disproportion is not just because of our big-brained babies. It is also because of our narrow pelvises, which are a consequence of the modern human form of bipedalism. Without going into too much detail, narrow pelvises are needed for efficient weight transfer during bipedalism. The human pelvis is a compromise between the need to have an efficient form of locomotion and the need to give birth to babies with the biggest brains possible. Because no other mammal has this combination of a large brain and a bipedally adapted pelvis, cephalopelvic disproportion does not occur in other mammals (except some small primates like macaques, but this is an effect of body size and not pelvic morphology per se).

Strangely enough, the major increase in brain size in the human lineage corresponds exactly with the development of the modern human pelvis, about 2 million years ago. With the appearance of *Homo erectus*, we see large brains and essentially modern pelvises compared with australopithecines and even early *Homo* (*Homo habilis*). But how could these two changes occur simultaneously if larger brains lead to birthing difficulties? The answer is that in *Homo erectus* a fundamental change occurred in the mother-fetus relationship. From this point on, humans began to give birth to less fully developed babies. Whereas the brain at birth in chimps is roughly one-half its adult volume, in humans it is barely one-third. This allows for the passage of a baby through a narrower birth canal, but it results in a baby that is virtually helpless for most of the first year of life. In my opinion, this adaptation may represent the single most important event in the evolution of our genus. Birthing helpless babies means more cooperation among individuals for the care and feeding or infants, which builds social bonds and dependencies. To catch up for having been born with less completely developed brains, human babies continue to grow their brains at fetal rates for the next year. But instead of this growth occurring in the relatively boring environment of the uterus, it occurs as they are getting to know their parents and other members of their community and while they are experiencing stimuli from the world outside the uterus. These social interactions and external stimuli actually have a direct impact on the developing brain, molding connections within the brain and making the complexity of human cognition possible. So something that sounds bad (cephalopelvic disproportion) is actually a side effect of something that was critical in the development of modern human biology and behavior. And the first direct evidence of significant increases in brain size in the primate fossil record is recorded in the late Miocene of Europe.

Another important clue about the place that *Rudapithecus* occupies in the evolutionary history of the great apes comes from the structure of the skull. All three *Rudapithecus* skulls have the same configuration of frontal sinuses and brow ridges that we saw in *Hispanopithecus*, which is strong evidence of a connection to African apes. In addition, the skull of Gabi shows how the face and the braincase were connected. In orangutans and hylobatids, when viewed from the side the face is tilted upward compared to the braincase. In African apes and humans it is tilted downward (figure 8.4).

An upwardly titled face, or airorhynchy, is fairly rare in primates, and among the apes it is most strongly developed in orangutans, as noted earlier. Both gibbons and siamangs and *Ekembo* have airorhynchy, but it is less pronounced than in *Pongo.* The downward tilt of the face in African apes and humans, known as klinorhynchy, is probably a more recent event that unites the hominines, because all the hominoid taxa that branched off before the hominines evolved (orangutans, hylobatids before that, and *Ekembo* even earlier), are upward-facing to some degree. It is therefore likely that hominines evolved from an airorhynch ancestor and developed klinorhynchy before each lineage of hominine separated. As we saw earlier, this type of character that unites a lineage and is not found in the next most closely related lineages is called a derived character (synapomorphy). True frontal sinuses and brow ridges are also hominine derived characters, and they may also be related to facial orientation, although we don't yet understand the mechanics. At any rate, *Rudapithecus* is the oldest specimen that we can reconstruct with confidence as being klinorhynch, and this is an important piece of the puzzle in reconstructing hominine origins.

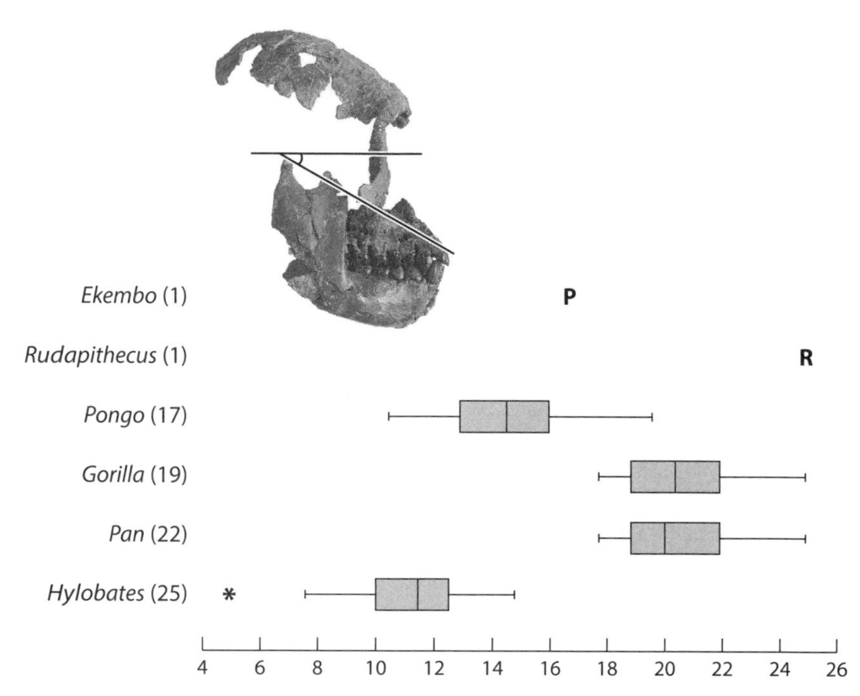

FIGURE 8.4. Facial orientation. The specimen is RUD 200 and its facial angle as measured from the line between the jaw joint and the bottom edge of the eye socket relative to a line from the jaw joint to the front tip of the upper jaw. This measurement gives a quantitative description of the tilt of the face, with higher numbers indicating more a clockwise tilt, and lower numbers a more counterclockwise tilt. *Rudapithecus* is at about 25 degrees (roughly 5 o'clock), while *Ekembo* is closer to about 16 degrees (about 3 o'clock). Orangs are earlier and African apes are at around 4:15, between *Ekembo* and *Rudapithecus*. Numbers (in degrees) refer to the angle indicated in the figure. Numbers in parentheses are sample sizes.

Sinuses that surround the nose, the paranasal sinuses, include the frontal sinuses as well as sinuses in the maxilla, ethmoid, and sphenoid, bones that surround the nose. Paranasal sinuses are present in most mammals but they are especially well developed in African apes. The frontal sinuses in particular are quite large in modern hominines, filling most of the brow ridges in African apes and a large portion of the front of the skull in humans. What is the function of these elaborate structures? There is no doubt that air spaces in the skull are lighter than solid bone, and a lighter skull requires less energy to support than a heavier one. So the development of paranasal sinuses may have something to do with making the skull lighter, which may be especially important if it already contains a large, heavy brain. Yet if that were the only reason for the development of the paranasal sinuses in African apes, why don't we see enlarged frontal sinuses in orangs, which have heads as heavy and brainy as chimps and gorillas?

Another interesting possibility has to do with the likelihood that paranasal sinuses function as resonating chambers. They are air filled and encased in solid bone, which vibrates as sounds are produced by the throat. As the skull vibrates, the paranasal chambers probably affect the quality of sound produced. Since they differ in the details of their size and shape from one individual to the next (like fingerprints), their resonating effects will produce different tones in each individual, which may be one reason that each of our voices is unique. Think about how your voice sounds when you have a cold and your sinuses are filled with fluid. With the resonating qualities of the sinuses altered, the sound of your voice changes. Having a unique, identifiable voice is important in social species in which each individual has a specific social role and identity, especially if you live in a dense forest and cannot always see the individual with whom you are communicating. Think of how convenient and easy it is to recognize the voice of someone you know on the phone. So the presence of variably sized and shaped, fairly extensive paranasal sinuses in African apes, which are much more social than orangutans, may serve an important social normative function by giving each and every individual a unique-sounding voice. This is consistent with the fact that chimpanzees, at least, have many different vocalizations that serve to communicate information about food, danger, and emotional state. I will discuss the possible significance of the brow ridge in chapter 9.

Another fascinating aspect of the skull of *Rudapithecus* is the inner ear. All mammals have structures in the inner ear that function to maintain a sense of balance in three dimensions. Deep in the temporal bone of the skull, an extremely hard structure (aptly named the petrous portion from the Latin for "rock") contains three semicircular canals, each oriented in a different plane. These canals contain fluid and sensors that communicate directly with the brain to give the individual a sense of its position in three dimensions. Because balance is critical for safe movement, the size of the canals and the diameter of their curvature strongly correlate with the way an animal moves. A great deal of research, initiated by my colleague Fred Spoor, seeks to quantify the dimensions of the semicircular canals and correlate them to different patterns of locomotion in primate evolution.

However, the only way to see the canals effectively, since they are embedded in the petrous bone, is with high-resolution, or micro, CT scans. My colleagues and I recently completed a study of the semicircular canals of the two skulls we have from Rudabánya using a micro CT scanner and the expertise of the researchers at the Max Planck Institute of Evolutionary Anthropology in Leipzig, Germany. The results show that *Rudapithecus* is most similar to highly arboreal, living, suspensory primates that need an extremely well developed sense of balance to safely move through the complex forest canopy.

Rudabánya has also yielded a large collection of limb bones, as we find at Can Llobateres. The femur that belongs to Gabi is smaller than the one from Can Llobateres but is similar in shape and strongly indicates this animal lived in the trees. Gabi is unique in that her pelvis is known. Like her femur, her hip is most like those of living highly arboreal apes, such as gibbons and orangutans. The hand and finger bones of Gabi and other members of her species, *Rudapithecus hungaricus*, show the same features, which are indicative of a suspensory ape. The elbow joint, with its specialized anatomy for maintaining stability in a wide range of positions (like the elbow of *Sivapithecus* described in an earlier chapter), has long been recognized as like that of a great ape, and even an African ape. *Rudapithecus* almost surely spent most of its time in the trees, and its enhanced grasping and climbing abilities made it able to exploit resources in any part of the tree canopy, from the main branches to the terminal stems.

So the dryopithecins, arriving in Europe between 12 and 13 million years ago, are closely related to the living African apes and to humans. They were not like any living ape but instead show a combination of features present in all living great apes.

OURANOPITHECUS

Toward the end of what one colleague called the "hominoid experiment in Europe" a new kind of ape appeared, one that looked more like our ancestors the australopithecines than any other Miocene ape. However, appearances can be deceptive. *Ouranopithecus* is known from a number of sites in Greece, near the modern cities of Thessaloniki and Athens, and possibly also from Turkey, Bulgaria, and Iran. It lived in a wooded setting, with more open spaces than other dryopithecins. It was larger and may have spent more time on the ground.

Ouranopithecus has been called the ancestor of our ancestors by my colleagues Louis de Bonis from France and George Koufos from Greece. De Bonis and Koufos have interpreted the similarities between australopithecines and *Ouranopithecus* to indicate that *Ouranopithecus* is ancestral to *Australopithecus*, and there are compelling arguments both for and against this idea.

Louis de Bonis and George Koufos point to similarities between australopithecines and *Ouranopithecus*, such as canine reduction; large, robust mandibles; and large, thickly enameled molars. There are also details of the jaw joint and premaxilla that they interpret as being shared with australopithecines. However, there is another explanation for the anatomical features shared between *Ouranopithecus* and early hominins.

I think of *Ouranopithecus* as a dryopithecin on performance-enhancing drugs. *Ouranopithecus* has the same set of features that link dryopithecins with African apes, including the beginning of brow ridges and a palate that closely resembles that of a gorilla. However, *Ouranopithecus* is much larger than any other dryopithecin, and it has dental characteristics that resemble those of australopithecines, our fossil ancestors. The molars are large and have very thick layers of enamel, broad flat cusps, and shallow basins, all of which are typical for a hard-object feeder, as described earlier.

Correspondingly, the jaws are large to absorb the stress of heavy chewing. In addition, the upper canines, which are long and dagger-like in most dryopithecin males, appear shorter in *Ouranopithecus* and somewhat resemble the canines of australopithecines (plate 14), though all the male upper canines are heavily worn, so this resemblance may be misleading. However, in our ancestors, both australopithecines and *Ardipithecus* (which I discuss in chapter 9), the canines are clearly reduced in males and wear in a different way.

I mentioned earlier that canine teeth reveal important attributes of social organization in anthropoid primates. Typically, males have large canines and females have much smaller canines. This dimorphism is generally thought to be associated with evolutionary selection from two directions. One involves intermale competition. Males display their canines in encounters with other males to obtain access to a variety of resources, to assert status in a given social situation, and to gain and restrict access to females. Therefore, there is selection for increased canine size in males to enhance their chances at the mating game. The size of the canines influences a male's ability to intimidate other males though a display of weaponry, a threat that usually serves to avoid a fight that could lead to injury on both sides. Fighting with those large canines is the nonhuman primate equivalent of the cold war insanity Mutually Assured Destruction, aptly referred to as MAD.

On the other hand, female choice plays a role that is probably at least as important or perhaps even more so. If females prefer to mate with males that have smaller canines, it does not matter how much competition among males selects for increases in canine size; the males with smaller canines will father more offspring and by definition be more fit from an evolutionary perspective. So females must prefer males with larger canines for canine dimorphism (differences in male and female canine length) to develop. It is probably not the canines per se that are attractive to female primates, although who knows? It is more likely the body size of larger males with big canines and their proven ability to access resources that are attractive to females, because the genes responsible for the characteristics of a dominant male will be transmitted to his offspring. Of course, all of this occurs unconsciously in both males and females, but in the end both forces contribute to this phenomenon.

When canine size diminishes in males, something else must be going on. This is comparatively rare in primates. As mentioned earlier, human males have reduced canines that are not noticeably different from those of females, although they are in fact larger on average. The common wisdom is that this size reduction reflects either a decrease in intermale competition or an increase in intermale cooperation and a concomitant preference among females for males with a lot of buddies. The unconscious calculation here is that males with small canines are likely to have cooperative relationships with other males that enhance their ability to access resources and perhaps even to assist females in defending her offspring against predators. So canine size is an important indicator of social organization in living primates. Some have suggested that *Ouranopithecus* males have smaller upper canines than *Dryopithecus*, *Rudapithecus*, or *Hispanopithecus*, although again these teeth are all very worn down. However their lower canines are not reduced in height, and both upper and lower canines in *Ouranopithecus* are still larger than in any of our ancestors. So, are the large, flat, thickly enameled molars, robust jaws, and possibly reduced upper canines of *Ouranopithecus* indications that it is an ancestor of our known ancestors, or is there another explanation?

Most researchers interpret the similarities between *Ouranopithecus* and australopithecines to be homoplasies (independent, parallel evolutionary developments, as described earlier). There are several reasons for this interpretation. All of the similarities between *Ouranopithecus* and australopithecines are concentrated in the jaws and teeth. As we saw in the story of *Ramapithecus*, it turns out that large jaws and teeth with thick layers of enamel have developed many times, not only in apes but also in other primates and even other mammals. As I mentioned before, *Ouranopithecus* lived in an environment that was drier and had less forest cover than the habitats of other dryopithecins. It was larger and probably foraged more frequently on the ground as well, although it may have retreated to the trees at night, as do most living primates. *Ouranopithecus* appears to have evolved jaws and teeth that independently resemble those of australopithecines because they were doing similar things. Both foraged on the ground and benefited from a powerfully built chewing apparatus that enabled them to exploit a wide range of resources. In both the forest and more open settings, many potential food items are simply not available to animals that cannot penetrate the various protective coverings, such as shells or tough husks, or to animals that cannot find or capture them.

So the adaptations of *Ouranopithecus* that resemble those of australopithecines evolved independently under similar circumstances. In many other features *Ouranopithecus* more closely resembles *Rudapithecus* than *Australopithecus*, especially in the palate and front teeth. Furthermore, chimpanzees and the earliest humans, such as *Ardipithecus*, resemble one another even more than *Ouranopithecus* looks like an australopithecine. *Ardipithecus* does not have the massive jaws and thickly enameled teeth of *Ouranopithecus* and *Australopithecus*, yet it was probably a biped, and nearly all researchers accept it as one of us, a hominin. All of the similarities in dental and palatal structure between early humans and chimpanzees would have had to evolve in parallel if *Ouranopithecus* was indeed the ancestor of our ancestors. And bipedalism would have had to evolve twice if *Ouranopithecus* were more closely related to australopithecines than to *Ardipithecus*. For these reasons, I interpret *Ouranopithecus* as a very interesting case study in what happens to a forest ape that moves out into more open country. It seems almost as if there is a contingent response whenever this occurs: bigger jaws; larger, more thickly enameled teeth; and possibly smaller but not necessarily shorter canines (probably to make room in the jaw for bigger molars). Other examples of this trend include *Sivapithecus* and *Gigantopithecus*, which were both discussed in chapter 7.

OREOPITHECUS

Ouranopithecus adds a lot to our understanding of the diversity and success of the great apes of Europe, but in terms of unusual morphology, this last ape is really off the charts. The words "enigma" and "enigmatic" are more often used in articles about *Oreopithecus* than about any other primate I know of. *Oreopithecus* is known from several sites in Tuscany and Sardinia, in Italy, and it has what is probably the most perplexing combination of characteristics of any fossil ape. *Oreopithecus* is most commonly associated with densely forested settings. It is found in coal deposits, for example, which form only under humid, swampy forest conditions. It was also an islander. Sardinia is still an island today, in the Mediterranean west of the Italian peninsula, but back when *Oreopithecus* lived, some 8 million years ago, Tuscany was also an island. It was not until more recently that it was joined to the mainland of Europe, a consequence of the rotation of the African continent discussed earlier and the formation of the Alps. The volcanism and earthquakes that still rock the Italian peninsula reflect that fact that these tectonic movements are ongoing.

The history of the interpretations of *Oreopithecus* is fascinating. The first specimen, a lower jaw, was described by the French paleontologist Paul Gervais in 1872. While recognizing the unusual morphology of the teeth, Gervais nevertheless placed it among the hominoids. Shortly thereafter other interpretations emerged, including comparisons with gibbons, Old World monkeys, and even fossil prosimians. William King Gregory, of the *Dryopithecus* Y-5 pattern fame, even suggested in his magnum opus of evolution, *Evolution Emerging*, that *Oreopithecus* might be closely related to the early ancestors of pigs.

By the end of the nineteenth century, a consensus was emerging that *Oreopithecus* was a highly specialized and unusual cercopithecoid. This idea, first proposed by another prominent paleontologist of the time, the German researcher Max Schlosser, was widely accepted until the work of Johannes Hürzeler. Starting in the late 1940s, Hürzeler began to publish detailed studies on the dentition of *Oreopithecus* and concluded that it was not only a hominoid but what we would call today a hominin, like *Australopithecus*.

Early on Hürzeler focused on the premolars of *Oreopithecus*, in which he saw affinities with humans. With the discovery and publication of a spectacular skeleton of *Oreopithecus* in 1960, Hürzeler extended his comparisons with hominins to include the face, which is short, and the pelvis, which is short and broad, as in hominins. Today all researchers who have studied the remains of *Oreopithecus* interpret it to be a hominoid of some sort. It is amazing that with many jaws and teeth and a virtually complete skeleton there still remains so much uncertainty about the exactly evolutionary position of *Oreopithecus*. Some would place it among the dryopithecins, but for reasons discussed below, I think this is very unlikely. To me, *Oreopithecus* is the end of a lineage that probably goes back to the early Miocene of Africa and retains many primitive features of the face, limbs and braincase.

One final note on the history of *Oreopithecus*. I had the pleasure of meeting Hürzeler in the late 1980s in Basel, where he spent most of his career, while I was working on my dissertation. Hürzeler was a giant of European vertebrate paleontology, like Kretzoi, Zapfe, and Crusafont. I was visiting the Natural History Museum in Basel to see fossil primates but not specifically *Oreopithecus*. My wife and our two little children were there as well, and they were planning on visiting the exhibits while I worked in the basement, where the collections are stored. We turned to the elevator, strollers in tow, and as the doors opened we saw an elderly, well-dressed gentleman, who was helpful directing us to the exhibits. He was very kind, asking us where we wanted to go and helping us with the strollers. He was surprised when I did not get off on the second floor, and I explained that I was here to study the fossils. That was when he introduced himself as Johannes Hürzeler. I thought he was the elevator attendant. He had been retired for many years and was the last person I expected to meet. We had a wonderful chat on the way to the collections and throughout the day (he was very pleased that I spoke French). At the end of the day he asked me if I was interested in seeing the famous *Oreopithecus* fossils. I said yes of course, and he showed me to the corner of the room where he had a small office. Under his desk was a small cabinet with a lock, which Hürzeler opened. He pulled out the top drawer and said, "This is *Oreopithecus*," and promptly shut the drawer. As frustrating as that might sound (I have since studied all the fossils), it is one of my favorite encounters with one of paleontology's luminaries of the last century.

So *Oreopithecus* evolved in isolation on one or more islands, without much competition from other animals and especially without predators. There were few animals such as big cats, large snakes, crocodiles, or other predators to bother *Oreopithecus*, so it evolved, unconstrained in a sense, in unusual directions for an ape. *Oreopithecus* has a very primitive-looking face that in many ways looks more like *Ekembo* than a European great ape. In many details of the bones of the skeleton *Oreopithecus* is also similar to *Ekembo*. However, *Oreopithecus* differs from *Ekembo* in major way: the former has greatly elongated arms that show many features related to an ability to suspend and move the body below branches, as described earlier. The details of the elbow, including the reduction of the olecranon process and the hinge-like morphology of the elbow joint, are also found in *Oreopithecus*. In my view, as *Oreopithecus* evolved in isolation in its Tuscan swamp, it grew in size, given that it had abundant food and few predators. As it got larger, it became more difficult for *Oreopithecus* to stay balanced walking on the tops of branches, so it began to move below them, which caused selection for suspensory characteristics. As with the chewing apparatus of *Ouranopithecus*, the suspensory features of *Oreopithecus* probably evolved in parallel, given a predisposition in many primates to hang below branches (plate 15).

While *Oreopithecus* and *Rudapithecus* were similar in body mass, *Oreopithecus* had a brain probably less than half the size of that of *Rudapithecus*. Reduction in brain size is fairly unusual in evolution, but *Oreopithecus* is the exception that proves the rule. With few if any predators and few competitors, *Oreopithecus* evolved adaptations to exploit abundant resources on the islands. In addition to becoming large and suspensory, *Oreopithecus* developed powerful jaws, very large chewing muscles, and incredibly cresty, cuspy teeth perfectly suited to slicing through very tough, highly fibrous foods. *Oreopithecus* has a set of teeth more strongly adapted to folivory than just about any other primate, living or extinct, according to many analyses. So *Oreopithecus* became the ape version of a tree sloth: large, suspensory, slow moving, and not especially clever. It is not much of an intellectual feat to move slowly among the branches gathering leaves and other abundant forest resources (someone once told me that they thought *Oreopithecus* may have eaten pine cones, which are known from the sites), especially if you do not have to worry about predators. So once again, apes evolved in a diversity of directions depending on the ecological circumstances.

As we will see, changes in the ecology of Europe starting about 10 million years ago led eventually to the extinction of the great apes of Europe and opened the door for the evolution of the modern subfamilies of the African apes and humans and the orangutan.

In this and previous chapters I try to tell the story of the origin and evolution of the first members of our subfamily, the Homininae, and how the most important characteristics of our lineage evolved in Europe and not in Africa. There is always a chance that fossils will be found in Africa that will effectively falsify the hypothesis that the subfamily we share with gorillas and chimpanzees spent some time in Europe and evolved such features as suspensory positional behavior and large brains there, but for now I find the fossil evidence from Europe the most convincing.

The beginning of the late Miocene was a glorious time for the great apes of Eurasia. They flourished from Spain to China, but as the climate of the supercontinent became more seasonal, with more pronounced dry seasons, cooler winters, and periods of fruit shortages, the apes of Eurasia declined in abundance. Eventually most of them went extinct, unable to survive the ecological transition. In my view, at least one population of great apes from Europe was able to disperse south, eventually into Africa, following the subtropical ecological zone. This population eventually became the ancestor of the African apes and humans. Once back in Africa, a dryopithecin of some sort (whether it was more like *Rudapithecus* or *Ouranopithecus* is unknown) successfully adapted to the less seasonal, subtropical climate of Africa and underwent a new adaptive radiation in the region.

In chapter 9 I will explore the implications of the fossil record of the apes for our understanding of the evolution of the human lineage. I will also discuss recently discovered specimens that offer a challenge to the out-of-Europe hypothesis.

CHAPTER 9

BACK TO AFRICA AGAIN

In the time that I have been studying ape and human evolution, we have moved from a simple, unilinear concept, one species leading to the next, to an incredibly complex picture. When I started learning about paleoanthropology as a first-year undergraduate in 1977, it seemed as if the researchers had it all figured out. We had *Ramapithecus*, the ancestor of humans, living in Kenya, Turkey, and South Asia, and *Dryopithecus*, the ancestor of the great apes, living in Europe, Africa, and Asia. Then there were *Australopithecus*, *Homo habilis*, *Homo erectus*, and *Homo sapiens*, evolving in a single direction, toward bigger brains and smaller faces. Fairly quickly and just in time to draw me in, things began to get interesting.

By the early 1980s, the animals from Europe, Asia, and Africa that had all been lumped into *Dryopithecus* reemerged as separate entities with different evolutionary histories. *Ramapithecus* was merged with *Sivapithecus*, which was and still is, according to most researchers, linked with orangutans. *Proconsul* was pulled out of *Dryopithecus* and recognized as either an early ape or having a link to African apes. We lost our direct link with the Miocene (*Ramapithecus*) at around the same time that the molecular biologists were telling us that the human lineage could not be 14 million years old, the age of *Ramapithecus* (or *Kenyapithecus*) in Kenya. *Ouranopithecus* was discovered in the 1970s, but better specimens described in the 1980s got the attention of many researchers who saw links with orangs or gorillas. *Dryopithecus*, once again limited to Europe, was left in the dust, largely viewed as a mildly interesting side branch of ape evolution. Almost no one was working on poor old *Dryopithecus* anymore, and I saw my opening.

Dryopithecus was the topic of my dissertation, which I completed in 1987. Unlike most researchers, I saw evidence of a connection between the great apes of the Miocene of Europe and living African apes and humans. It would be a long time, however, before I could convince many people that this is a plausible hypothesis, not to mention a likely scenario. In the last two decades of the twentieth century, many new genera and species were identified, some from previously known fossils, but mostly from new discoveries. The same was happening with Plio-Pleistocene hominins (australopithecines and early *Homo*). The simple straight line of ape and human evolution was rapidly becoming a big, incomprehensible mess.

While some throw up their hands in disgust and evoke parallel evolution, claiming that we will never make much progress in our understanding of ape and human evolutionary history, most of us, fortunately, have decided to push ahead. While there is a great deal of parallel evolution, there are patterns. *Ekembo* is primitive and, simply put, part monkey and part ape, at least morphologically (it is all ape phylogenetically). Middle Miocene apes are more modern looking, especially in the teeth. Late Miocene apes have many features in their heads and in their trunks and limbs in common with living apes. There are great challenges in trying to make sense of the fossil record because evolution is inherently messy. Animals are constantly confronted with demanding environments and are constantly changing. Experiments abound, and animals are often evolving similar-looking adaptations. Different parts of the body are evolving at different rates and different times. I would go so far as to say that it is antievolutionary to imagine a straight line in which the few fossil species that we can discover happen to be those that led directly from our monkey-like ancestors to modern apes and humans. Today we are much closer to seeing the true picture, the real complexity of ape and human evolution, than we have ever been.

LEAVING EUROPE BEHIND

Eurasian great apes appear to have been driven from their ancestral homelands in Europe, South Asia, and East and Southeast Asia by global climate change. Most of the world is very concerned about global climate change or global warming, but most people do not realize that dramatic changes in climate have occurred thousands, if not millions, of times in the course of the history of the Earth. I prefer the term global climate change to global warming because "warming" is an oversimplification. The effects of global climate change include cooling in some areas, warming in others, and changes in precipitation patterns resulting in droughts and flooding. The long history of global climate change before humans evolved does not minimize the importance of modern global climate change, which is caused in large part by humans and is occurring at an accelerated and dangerous rate, but global climate change is not new. Ironically, what is posing a huge challenge for future generations of humans, global climate change, is what may have sparked the origin of humans in the first place.

No one really knows exactly why climate has changed on a global scale so frequently and dramatically during the history of life on earth. Part of it has to do with the process known as plate tectonics. The continents, which are generally made of light rocks, essentially float on the denser oceanic crust and have moved around all over the globe during billions of years. Part of the process involves the creation of new ocean crust, as magma moves up from deep in the earth, essentially spreading the seafloor and pushing the continents around. As continental crust squeezes up against oceanic crust, it tends to get thrust downward, toward the earth's core, resulting in melting of the relatively lighter rocks that compose the continents. This downward forcing, or subduction, of the continents creates volatile chambers of molten rock that, with enough pressure, explode, producing volcanoes and other tectonics-related events. Most volcanoes are fairly localized, but there have been supervolcanic eruptions in the past that have significantly modified the chemical composition of the atmosphere (especially with regard to carbon dioxide), which paleoclimatologists consider to have been significant in altering the earth's climate. Supervolcanoes still exist. The one that I am most aware of is the volcano under Yellowstone National Park, which is responsible for the spectacular geological formations there. No one knows what will happen when that volcano awakens.

Plate tectonics leads, of course, to changes in the number and position of the continents. When continents are closer to the equator than to the poles, life adapts to warm climates with little seasonal variation, and the opposite is the case when portions of the continents are closer to the poles. The position of continents affects the configuration of the oceans, since that is how we essentially define them (the Atlantic and Pacific Oceans, for example, are separated by the Americas). Ocean size and shape has effects on ocean currents, which have profound effects on land masses. Consider, for example, the Gulf Stream, a powerful, warm ocean current. Passing just off the coast of Cornwall, in southwestern England, it gives the Cornish their very temperate, if not subtropical, climate, the envy of the rest of Great Britain. Finally, as plates collide with one another (such as the collision of Europe and Africa in the early Miocene), mountains are formed, which have both local and continent-wide effects on climate, producing such phenomena as monsoons and rain shadows.

Other factors that affect climate include the tilt of the earth's axis (the virtual line around which the earth rotates) relative to the plane of its revolution around the sun, as well as variations in the energy output of the sun. These and other factors (meteor impacts, for example) have had profound effects on the global climate for millions of years before a single anthropogenic carbon emission was made. Once again, this is not to minimize the effects of carbon emissions on the current phase of global climate change. There is overwhelming scientific evidence that we have brought this situation on ourselves and that we have dangerously accelerated an otherwise natural process.

Before looking at the evidence for hominines in Africa from 12.5 million years onward, let's have a look at the evidence for that global climate change I was talking about, the one that may have, by random chance of course, contributed to human origins.

The world was undergoing profound changes in the late Miocene. In fact, these changes start back in the middle Miocene, after a period known as the middle Miocene climatic optimum. How do we know that the climate was like millions of years ago? Researchers use what are called proxies to reconstruct the ancient climate, or paleoclimate. Proxies from the recent past include the analysis of ice cores extracted from the Arctic and Antarctic, which provides information on climate via the chemical composition of water going back thousands of years. But if you want to know about climate in the more distant past, you need to use less direct methods, which include assessments of the climate preferences of fossil plants and animals based on analogies to similar animals alive today. Pine trees and willows, which can be identified by their pollen as well as leaf impressions, tell paleobotanists a lot about the local ecology. Pine trees tend to be found in drier, cooler climates, whereas willows are often found near a source of water. Animals—such as zebra, which live in open country and , flying squirrels, which are forest dwelling—are indicative of local ecological conditions. Researchers also use a technique known as ecomorphology, in which the taxon is ignored (a so-called taxon-free approach) and only the morphology is analyzed. Leaf structure, for example, is very revealing of relative humidity and seasonality regardless of the species. Limb length and especially dental morphology reveal aspects of adaptation correlated with ecological factors. For example, tall teeth indicate grasslands, since eating grass tends to wear teeth down. (Remember what I said about looking a gift horse in the mouth.) Indications of arboreality necessarily imply the presence of trees. The structure of snail shells and the tightness of their coils are indicative of salinity and temperature. Finally, chemical analysis of fossil soils (paleosols) and fossil plants and animals can reveal a great deal about paleoclimate. Differing ratios of various isotopes of carbon, oxygen, nitrogen, strontium, and other elements provide indicators of temperature and precipitation. One of my students, Laura Eastham, is finishing up an exciting analysis of the geochemistry of Rudabánya. She has found that in contrast to our previous assessment, Rudabánya was probably more seasonal than we thought, which must have presented a special challenge to the primates that lived there. Researchers can also reconstruct the photosynthetic pathways in different types of plants, which differ between grasses and most forest plants, from the chemistry of the soils in which they grew. This is one of the major sources of information about the decline of forest cover and the spread of grasslands at the end of the Miocene.

The Miocene was a time when global temperatures were significantly higher than today, and it is one of the factors that allowed apes to disperse into Europe from Africa. After the middle Miocene climatic optimum, which lasted from about 15 to 17 million years ago, a cooling and drying trend began. At first, apes in Europe adapted, which may also account for such evolutionary innovations as orthogrady and suspension. It may even account for the increase in brain size and changes in growth and development compared with the early Miocene apes. But by 9 million years ago, the climate in Europe had simply become too seasonal, cool, and dry for apes to survive. For the most part, apes rely on a steady, yearlong supply of fruits that are not present in temperate zones with cold winters. What was bad for the apes was good for the monkeys. They appeared in Europe around 11 million years ago but were rare until receding forests and increasing seasonality forced the apes out. By about 8 million years ago, monkeys such as *Mesopithecus*, an extinct Old World monkey related to living leaf-eating monkeys (colobines), were widespread in Europe. Later they would be replaced by the ancestors of living macaques and relatives of the baboons. There is only one site in Europe that may provide evidence that monkeys and apes lived together in the late Miocene, whereas monkeys and apes are commonly found in the same forests today. For the most part, the monkeys moved in after the apes had already moved out. They were not in direct competition with one another, which was once proposed as a theory to explain the extinction of the apes of Europe.

The last apes to survive the cooling and drying of Europe were the Greek ape *Ouranopithecus*, with its thickly enameled teeth, and similar taxa from Bulgaria, Turkey, and Iran. All of these apes were better able to adapt to the grassy, sparsely forested conditions than the dryopiths, with their thinly enameled teeth. The exception that proves the rule is *Oreopithecus*, the swamp ape from Tuscany, which survived until possibly as recently as 7 million years ago only because it lived in splendid isolation in a relic forest on what was then an island having few competitors and no predators.

While apes were thriving in Europe and Asia in the late middle and early late Miocene, they were extremely rare in Africa. But each time one is found, it is hailed as proof that hominines evolved in Africa and that the Eurasian apes are just side branches. Let's have a look at the evidence for apes in Africa that are contemporaneous with what I call the Miocene hominines of Europe.

In chapters 7 and 8 I discussed a large number of fossils that have many similarities to living African apes and humans, the hominines. They range in date from about 9.5 to 12.5 million years ago and have hominine characteristics of the teeth, skull, and limbs. Some have suggested that all of these features evolved independently in European apes, but I doubt it. In addition, there are almost no fossils in Africa from the same time period that are more similar to hominines. Actually, there are no fossils from Africa from the same time period that can be reliably attributed to the African ape and human clade. However, by about 7 million years ago, a number of fossils appear offering stronger evidence that they are not only hominine (African apes and humans) but in fact hominin (humans and our fossil relatives).

QUESTIONABLE HOMININES

Specimens from Ngorora

Twelve and a half million years ago, after their heyday in the early Miocene, hominoids were still hanging on in Africa. Scientists have reconstructed one site, called Ngorora (pronounced "gorora"), to have been a forested environment rich in many different mammals. It should be a perfect place to find apes. There are apes there, but they are extremely rare: four teeth and a toe bone, to be exact. Because the time period is so poorly represented in Africa, the teeth have been analyzed several times, but with inconclusive results. A premolar and a molar were described more than twenty years ago by my colleagues Andrew Hill and Steve Ward. The upper molar has a partial cingulum and the premolar (a lower second premolar, or P_4), is *Ekembo*-like in that the front part (trigonid) has two tall cusps while the back part (talonid) is a low basin. In more modern apes, both parts of the tooth are closer to the same height. I recently found the toe bone in a drawer at the National Museums of Kenya, in Nairobi, that contained Miocene monkey teeth, but the toe bone in question is definitely from an ape. It is a small bone, and finger and toe bones are often misidentified, but this one is long and curved and has those telltale signs of powerful grasping

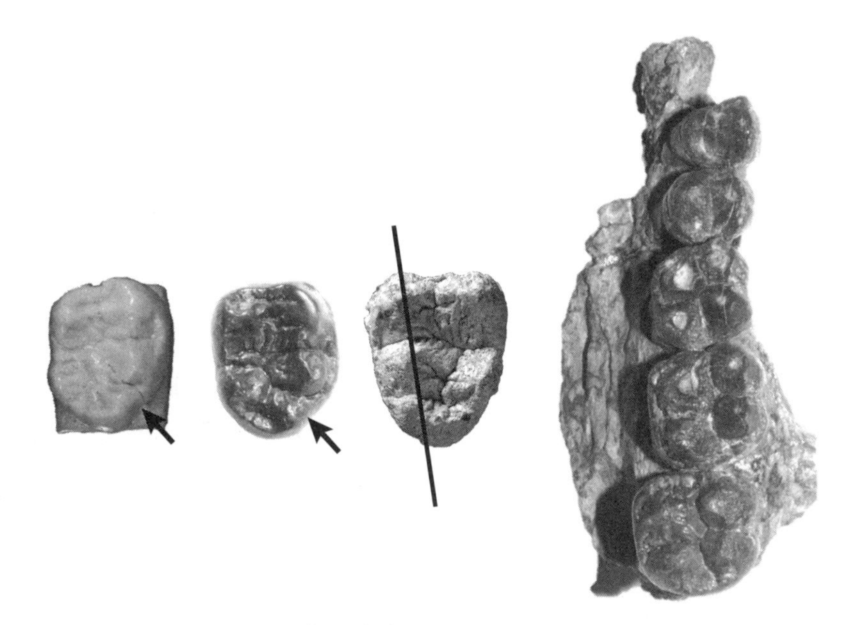

FIGURE 9.1. Composite of teeth from Ngorora, Nakali, Chorora, and Samburu. The lower right molar from Ngorora (*left*) and a lower right molar of *Nakalipithecus* (*second from left*) are very similar in morphology. Note the notch between the second and third cusps on the cheek side (*arrows*). A lower left molar from *Chororapithecus* has a different morphology (*second from right*). Note the alignment of the cheek-side cusps (*line*). Upper teeth chewing surfaces of *Samburupithecus* (*right*) have deep grooves between the well-defined cusps and the strongly developed lingual cingula. (Image second from left courtesy of Yutaka Kunimatsu. Image second from right courtesy of Gen Suwa. Other images by author.)

(flexor) muscles. When I had another look at the Ngorora teeth, I found that they most closely resemble a recently described species, *Nakalipithecus nakayamai*, which I will discuss below. As a preview, I can say that if *Nakalipithecus* is really 12.5 million years old and if it is a hominine, it may test my main hypothesis that hominines originated in Europe and dispersed back into Africa (figure 9.1).

Two other isolated discoveries have been described more recently by my colleague Martin Pickford and associates. They include a lower molar from Ngorora and a very poorly preserved mandible fragment from Niger. The lower molar is like the other teeth from Ngorora in that it resembles teeth from *Ekembo* and *Nakalipithecus* more than anything else. For example, it has a well-developed cheek cingulum, one of the ridges on the cheek side of lower molars and the tongue side of upper molars. It does not resemble late Miocene ape teeth from Europe. The Niger specimen is a piece of jawbone with a molar that is sliced in half. In my opinion, it is completely undiagnostic. The age of the specimen can only be narrowed down to between 11 and 8 million years ago because it is not clear exactly where it came from. Pickford and his colleagues have described other fossil material from the late Miocene of Africa that they have interpreted to represent taxa closely related to European fossil apes, chimpanzees, gorillas, and humans. Very few researchers in the paleoanthropological community accept these conclusions, mainly because the samples are so fragmentary.

Nakalipithecus nakayamai

This animal is known from a piece of lower jaw and a few isolated teeth from central Kenya, more specifically a place known as the Tugen Hills. Ngorora is also in the Tugen Hills but about 80 kilometers away. *Nakalipithecus* is 9.8 million years old and displays a mix of traits seen in both early and late Miocene apes. On the primitive side, *Nakalipithecus* retains cingula. The thick enamel recalls middle and late Miocene apes such as *Kenyapithecus*, *Sivapithecus*, and *Ouranopithecus*. One feature that is quite distinctive about *Nakalipithecus* is the morphology of the female upper canine. Canines, both male and female, usually have only one cusp, but this canine has a strongly developed bulge, or projection, on the tongue side, which scientists call a cuspule. I have yet to see another example of this, but the species that comes closest to having a similar morphology is *Ouranopithecus*. In fact, my colleagues who discovered and described *Nakalipithecus*, Yutaka Kunimatsu and Masoto Nakatsukasa, whom we met earlier in the discussion about *Nacholapithecus*, concluded that there is a good chance that *Nakalipithecus* is the ancestor of *Ouranopithecus*. They infer from this that *Ouranopithecus* dispersed into Europe after

Nakalipithecus lived and did not evolve from a dryopith, as I have hypothesized. However, *Nakalipithecus* is only a few hundred thousand years older than *Ouranopithecus*, and we have no way of knowing if older *Ouranopithecus* species remain to be discovered in Europe. In paleontological terms, they are essentially the same age. In addition, the canine cuspule on *Nakalipithecus* is more strongly developed than in *Ouranopithecus*. One could argue that it represents the derived condition, having evolved from *Ouranopithecus* rather than the other way around. The bottom line is that we need more fossils to decide between these competing hypotheses; however, because there are so many features of European great apes shared with hominines, my money is still on the European origins hypothesis.

Chororapithecus

Another African late Miocene potential hominine is *Chororapithecus*, from a 10.5-million-year-old site called Chorora in Ethiopia. *Chororapithecus* is known from nine isolated teeth—some in pieces, many with corroded enamel surfaces—making morphological comparisons difficult. My colleague Gen Suwa from the University of Tokyo found a way around this problem by using micro CT scanning, which reveals the pristine dentine surface under the damaged enamel. Suwa and his colleagues were able to identify a crest on the dentine surface that they say corresponds to a crest on the enamel surface of gorilla teeth, though the enamel crest is not found in *Chororapithecus*. They suggest that this crest represents an important adaptation of gorillas for processing fibrous foods, which they do with greater frequency than other apes. The teeth of *Chororapithecus* are also quite large, within the size range of those of living gorillas. Suwa and colleagues therefore conclude that *Chororapithecus* may be an early member of the gorilla clade. The problem is that if the dentine crest has the same evolutionary origin as the enamel crest in gorillas, it could not have evolved as part of a feeding adaptation because it is below the enamel and would never come into contact with food. The absence of a corresponding crest on the enamel in *Chororapithecus* means that it lacked the gorilla's adaptation for cutting through food fibers, casting doubt on the idea that its dentine crest has anything to do with the gorilla's enamel crest. If it were an adaptation to processing fibrous foods, I would expect the opposite pattern, that is, an incipient crest on the enamel surface that eventually becomes reinforced by a corresponding underlying dentine crest. The two mostly complete upper molars and the one intact lower last molar crown do appear to retain cingula, which is a primitive character. They also have thick enamel, in contrast to the thin enamel of gorillas. Their cusps are much lower and more rounded, and their basins much shallower than in gorillas. Once again, eight and a half or so teeth are just inadequate; we can't be sure what kind of animal *Chororapithecus* really is. To their credit, Suwa and colleagues characterize the evidence for a link to gorillas as inconclusive, even though they prefer this hypothesis. As they note, if *Chororapithecus* is in the gorilla clade, this would move the divergence of gorillas from the chimpanzee-human clade back to at least 11 million years ago, which again is much earlier than the consensus date of divergence of about 9 million years. However, a paper published in 2013 reports results from the lab of my colleague Todd Disotell of New York University that puts the gorilla-chimp/human divergence back to about 10.5 million years ago. They also push back the orangutan-hominine divergence back to about 17.5 million years. In my view both of these dates are realistic and broadly consistent with the fossil evidence. Nevertheless, the morphology of *Chororapithecus* does not allow us to conclude that it represents an early gorilla. We need more fossils. Here is hoping that all of my colleagues from Japan and elsewhere continue to find exciting new fossils in East Africa so that we can determine what these tantalizing fossil scraps mean.

Samburupithecus

The last potential hominine from East Africa is found back in Kenya, a bit to the east of the Tugen Hills at a place called Samburu Hills. The Samburu Hills hominoid is 9.5 million years old, making it the youngest in this first group of African potential hominines. The locality is thought to represent a wooded environment, though there is evidence of more-open country as well. This hominoid has been attributed to a new genus, *Samburupithecus*. Pickford and colleagues interpret this specimen—an upper jaw with the premolars and molars intact—to be most closely related to australopithecines. The teeth of *Samburupithecus* are a bit odd. Its cusps are tall and separated by deep fissures and have very strongly developed cingula, making them essentially identical with the teeth of early Miocene hominoids like *Ekembo*. *Samburupithecus* also has a low cheekbone, positioned close to the teeth and a narrow nasal aperture, again just like *Ekembo*. Using high-resolution CT scans of the palate of *Samburupithecus*, my colleagues and I have been able to virtually remove the matrix or sediment filling portions of the specimen, revealing a structure of the front part of the palate very similar to that of *Ekembo*. We have no bones below the neck for *Samburupithecus*, so we cannot conclude anything about how it might have moved around. While I am not clear on exactly what *Chororapithecus* and *Nakalipithecus* are, I am quite confident that *Samburupithecus* is the last of the proconsuloids that we know of, a relic from the glory days of early Miocene ape diversity in East Africa.

A common complaint from those who reject the back-to-Africa hypothesis is that the fossil record in the late Miocene in Africa is poor compared with the early and middle Miocene and that we have yet to find convincing evidence of the ancestors of crown hominids (great apes and humans) or hominines (African apes and humans) there because the region is so vast. Often this rationale is accompanied by the refrain, "absence of evidence is not evidence of absence." This is true. Africa is a huge continent and we have barely scratched the surface in looking for fossils of any type. Of course, Eurasia is pretty big as well, and Europe has a distinct disadvantage for paleontologists that a lot of Africa does not have: most of Europe is heavily vegetated and populated by people, cities, cropland, forests, parking lots, shopping malls, etc., which tend to hide fossil-bearing rocks. And the fossil record in Africa is not as poor as all that, as we have seen. In addition to the three sites described earlier, there are quite a number of localities in the 7-to-12-million-year range throughout Africa, many very rich in fossils and having a wide diversity of different types of mammals. Being forested, many of these localities would seem to be excellent places for hominoids to have lived, but the fact is that up until now not a single specimen with convincing evidence of crown-hominid affinities has been found.

Besides this, I question the wisdom of rejecting a hypothesis out of the expectation that it will be falsified when the appropriate evidence is eventually found. This is like convicting someone of a crime essentially because you are sure they did it, even if you cannot prove it but you are sure that you will eventually come up with some evidence. That's not how it works in science. We have to go with the evidence that we have, and the evidence indicates that in the late Miocene there were many species of hominids in Eurasia and very few if any in Africa. This does not prove that crown hominids originated in Eurasia, but it provides the basis for that hypothesis. This is a hypothesis that would be falsifiable by the discovery of crown hominids in Africa that are older than those we find in Eurasia. Until then, the evidence points to a Eurasian origin.

Aside from the fact that late Miocene hominids from Africa remain to be found, what other explanation would account for the presence of apes with clear crown-hominid affinities in Eurasia? In particular, if the crown hominines did not originate in Europe but evolved in Africa, why are there hominines in Europe starting at about 12.5 million years ago? Could these all be side branches of an earlier crown hominine population in Africa that awaits discovery? Talk about your complicated hypotheses, not to mention special pleading.

While the apes from Africa between 9.5 and 12.5 million years ago fill a time gap between middle Miocene apes and hominins, they do not fill an evolutionary gap. There is no good evidence that they are hominines or hominins, and they may in fact all be relics from the past, like *Samburupithecus*. However, by 7 million years ago or so, a new kind of ape came on the scene.

THE FIRST HOMININS

Sahelanthropus tchadensis

It is interesting that hominins do not appear before 7 million years ago, but afterward, they begin to appear in number. It does suggest that the fossil record is reliable and that hominins or hominines were not present in Africa earlier. At any rate, *Sahelanthropus tchadensis* is about 6 to 7 million years old, though the describers prefer the older date. Unlike earlier African Miocene apes, *Sahelanthropus* is from Chad, in western Africa, on the southern edge of the Sahara desert. It was unearthed between 2001 and 2002 at a place called Toros-Menalla by a team from Chad and France. At the time that *Sahelanthropus* lived there, this part of Chad was not a desert but a mixed environment bordering a large lake. Fossils of hippos, fish, amphibians, and crocodiles attest to the presence of fresh water. Monkeys and other arboreal mammals attest to the presence of forest. Zebras and antelopes, along with some grassland rodents like hamsters, indicate the presence locally of more open, sparsely forested conditions. The mosaic nature of the locality may in part explain the morphology of *Sahelanthropus*. It has been suggested that the presence of multiple ecological zones near the lake may have selected for a form of locomotion in which the animals were as adept in the trees as on the ground. In other words, at Toros-Menalla there may have been selection for an early form of bipedalism that permitted efficient movement across open country but with the retention of traits allowing for proficiency in climbing.

My colleague Michel Brunet, who described *Sahelanthropus* with his colleagues, picked a name with a message. It emphasizes their view that *Sahelanthropus* is a human (anthropus) and not an ape (pithecus). As I did with the skull of *Rudapithecus* we discovered at Rudabánya (I called her Gábi, after Gábor Herniak), Brunet gave a pet name to his discovery, which is also revealing of his ideas about the specimen. He called the specimen Toumaï, which means "hope of life" in Dazaga, the local language. There is some good evidence that Toumaï represents a hominin. Toumaï, the best specimen of *Sahelanthropus*, is a cranium (the skull minus the lower jaw). Although it is somewhat distorted, it is relatively complete, and it has been possible to correct the distortion through computer modeling. When this is done we can see what the bottom of the cranium, the basicranium, looked like (figure 9.2).

In mammals, the basicranium is a very crowded place. It is where the head attaches to the neck and is crammed full of muscles as well as all of the nerves, including the spinal cord, that extend from the

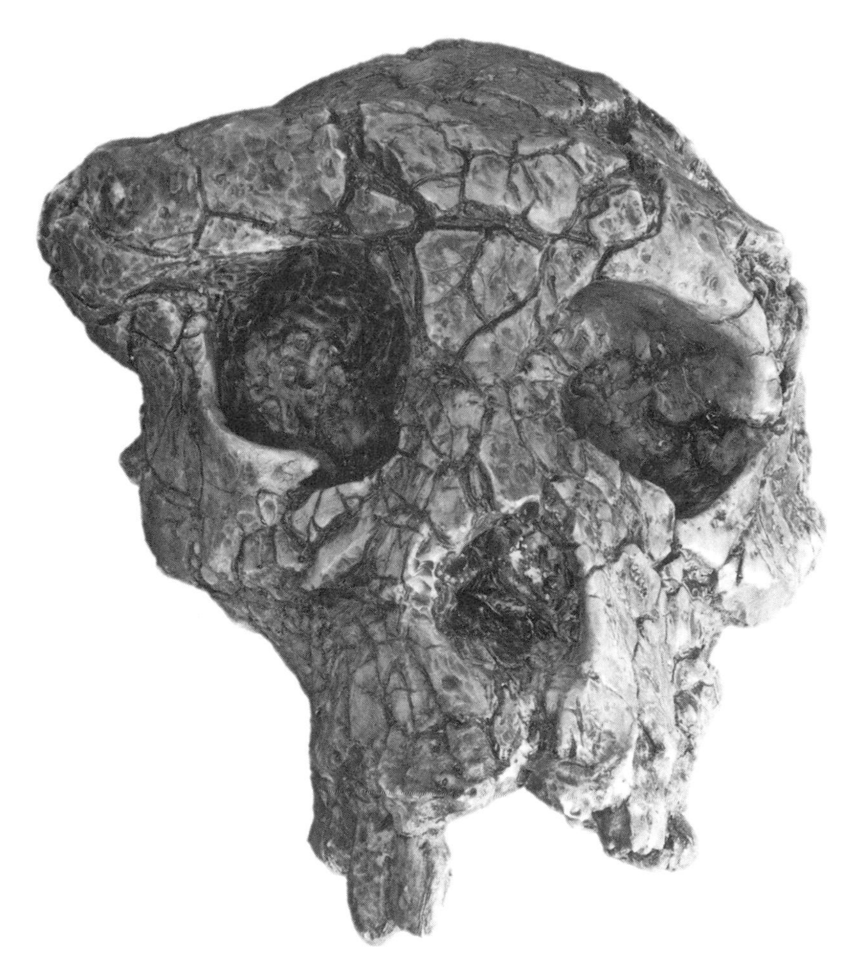

FIGURE 9.2. *Sahelanthropus* (a cast). Note the large brow ridges and the relatively flat face. (Image by author.)

brain. It has the jaw joint and the middle and inner ears and places for the attachment of the muscles, glands, and other soft tissue structures of the throat. And of course, the lower jaw attaches to the basicranium. One of the features of the basicranium that is unique in humans is the position of the opening that allows the spinal cord to connect to the brain, the foramen magnum. The "big hole" (literally translated from the Latin) is the opening to the braincase from which the spinal cord exits the head. It is actually filled with the lowermost part of the brain, the medulla oblongata, to which the spinal cord connects. In humans, because we are bipedal and need to have our heads more or less balanced on our necks, the neck attaches to the bottom of the head centrally, so that the front and back of the head are fairly balanced on the neck. It is not completely balanced, the front being a bit heavier than the back, but almost. (This is the reason that my students, when they fall asleep during my lectures, tend to drop their heads forward rather than backward.) Where the neck attaches is where the spinal cord exits the head, since it runs down the spine. In apes and all other primates, the foramen magnum is positioned toward the back of the skull because the head is placed in front of the body. When you have a foramen magnum under the skull, close to the middle, it is a good indication that the skull was balanced on top of the spine, which means an upright posture and bipedalism. In Toumaï the foramen magnum is placed near the middle of the bottom of the cranium, suggesting that that neck was placed below the skull and balanced, as in a biped. Also, in Toumaï the plane of the foramen magnum, that is, how it faces, is downward, as in bipedal hominins (see plate 16).

One curious thing about Toumaï is that the back of the braincase, behind the foramen magnum, is expanded, leaving lots of room for large nuchal, or neck, muscles. You may have noticed your own nuchal muscles when they become sore when you are tired or have been stooped over a computer for hours. These muscles are normally large when there is a heavy load in front of the neck joint, as in great apes, especially gorillas. They are needed to keep the head horizontal so that the animal can look forward. Humans have a small nuchal area because we do not need much muscle force to keep our heads horizontal, since they are already more or less balanced. Toumaï is about the size of a small chimpanzee. So it is odd that in Toumaï the nuchal area is expanded with room for a lot of nuchal musculature. Normally this area is for the attachment of muscles that hold the head horizontal, but if the head is already balanced in Toumaï, why is the nuchal region so large? It may reflect Toumaï's diet. Part of detaching food from its source involves movements of the head, such as tugging food away from the source, and nuchal muscles could be useful for this purpose. But other researchers have suggested that the large area for the attachment of the nuchal muscles indicates that the head was not balanced on the neck in Toumaï. It is worth mentioning that the conclusion that Toumaï had a centrally positioned and downward facing foramen magnum is based in part on a virtual reconstruction of the fossil, which inevitably comes with some speculation. In my opinion the nuchal area in Toumaï remains a mystery, but I do think it probable that Toumaï was a biped of some sort.

There is another thing that is unexpected in Toumaï. That is the development of the brow ridge. Toumaï's brow ridge is very pronounced. If you measure the thickness of the brow ridge from top to bottom, the brow ridge is as big in Toumaï as it is in a male gorilla, which is very unusual in a hominin, especially one the overall size of a small chimpanzee. We do not what role the brow ridge plays in primates. Researchers have attached stress sensors to the brow ridges in monkeys and had them chew various foods, finding that there are no strains transmitted to the brow ridge. So if it is not functional, at least not in the food-processing sense, what is the brow ridge for? Believe it or not, some researchers have suggested that it is a rain shield, or to protect the eyes from blows from a club! These explanations do not make a lot of sense given the fact that many primates, not to mention modern humans, endure rainstorms without brow ridges or without needing to ward off blows from a club.

One thing that is true about the brow ridge in primates is that when present, it is more strongly developed in males than in females. I like the idea that they are a product of sexual selection, like the antlers of elk or the horns of antelopes. For some reason female hominins may have preferred males with big brow ridges. In this case, bigger is better. Perhaps larger brow ridges are a side effect of higher levels of testosterone or other hormones that may confer more strength or endurance, desirable qualities to transmit to offspring. Unconsciously, females may have been favoring males with big brow ridges because of the survival attributes their offspring would inherit.

If I am right that the brow ridges in Toumaï result from sexual selection, it stands to reason that Toumaï is male. This is important because the canines in Toumaï are small compared with those of male great apes. In great apes females have smaller canines than males, probably also as a result of sexual selection. If the canines in Toumaï are small but male, then this is evidence of canine reduction. As I said repeatedly in the discussion of *Ramapithecus*, canine reduction is an important diagnostic character of hominins.

My colleague Mike Plavcan has spent many years documenting the correlation between canine dimorphism and social organization. He has shown that high levels of competition between males results in large male canines, that is, a lot of canine dimorphism. If a species lacks canine dimorphism, or if it is reduced, this indicates an absence of severe male competition, though it does not always correlate with increased cooperation. It can though, and in this case I think it probably did. Cooperation rather than competition implies higher levels of social complexity, which certainly characterizes us compared with the great apes. Toumaï may provide the earliest evidence of the enhanced form of social behavior that we see in modern humans today.

Toumaï has a relatively forward positioned and downwardly oriented foramen magnum, suggestive of bipedal posture, but a large nuchal attachment site, suggestive of a specialized way of obtaining food from the environment. He had small canines, compared with living apes, but very large brow ridges and lived in an environment that may have offered multiple challenges and opportunities, from forest to desert. While the jury is still out on the hominin status of Toumaï, in my view it is an early, if not the earliest, hominin.

Orrorin tugenensis

Orrorin tugenensis is another candidate for hominin status in the latest Miocene of Africa. *Orrorin* is from a site in the Tugen Hills of Kenya called Lukeino. Paleoecologists have reconstructed Lukeino as a mosaic environment, but one more thickly forested than Toros Menalla. Monkeys are abundant in the list of animals from Lukeino, indicating the presence of a forest. Unlike *Sahelantropus*, the most informative fossils of *Orrorin* are postcranial. Most of a femur is known, from the hip joint to probably about three-quarters of the way down. Sadly, the most informative part of the thigh bone, the lower end, the part that makes up the top part of the knee joint, is not preserved. In humans, the knee joint is absolutely unique due to the biomechanical requirements of bipedalism. It has adapted to support the body mass on one leg at a time, which is the way we walk, and the resulting anatomy is unmistakable. The top part of the thigh bone is less diagnostic, but nevertheless, analysis indicates that it most likely belonged to a biped.

My colleagues Brigitte Senut and Martin Pickford, who work at the Muséum National d'Histoire Naturelle in Paris, have described *Orrorin* as a biped more closely related to humans than to the group of hominins usually said to be the ancestors of *Homo*, the australopithecines. I think it is fair to say that they are the only researchers who favor this interpretation. There are differences in the femora of *Orrorin* and *Australopithecus*, such as the position of a bump called the lesser trochanter, but I don't consider this strong evidence that *Orrorin* is more *Homo*-like than *Australopithecus*. The position and development of the thigh musculature in *Orrorin* is similar to that we see in *Australopithecus* and strongly indicates that both animals were bipedal. The most thorough analysis of the *Orrorin* femur to date, by Brian Richmond and Bill Jungers, of the American Museum of Natural History and SUNY Stony Brook respectively, conclude that *Orrorin* is most like australopithecines. Another analysis headed by Sergio Almécija of George Washington University and the Universitat Autònoma de Barcelona concludes that the *Orrorin* femur also resembles that of Miocene apes, so it seems to be an intermediate form.

A brief aside before I continue with *Orrorin*. The Muséum National d'Histoire Naturelle is one of my favorite places in the world. First of all, it is in Paris. Besides that, it employed some of my heroes: Buffon, Lamarck, and Cuvier. Statues of all three grace the grounds of the Jardin des Plantes, where the museum is located. Lamarck's statue is especially touching. On the back there is a bas-relief with the great evolutionary biologist (he coined the term "biology") and his daughter. Sculpted in the bronze is the following; "La posterité vous admirera, elle vous vengera, mon père," which roughly translated is, "History will admire and avenge you, my father." Lamarck was criticized, in fact ridiculed, in his day for proposing that organisms respond to ecological conditions by changing over time; that is, they evolve. The mechanism he proposed to account for the transformation of species, the inheritance of acquired characteristics, was incorrect, but the ideas of adaptation and transformation were spot on, fifty years before Darwin. Buffon, the professor, and his students, Lamarck and Cuvier, were amazing and highly prolific scientists, and I admit to feeling their presence when I am in the museum and on the grounds.

Other bones attributed to *Orrorin* include parts of the humerus and a finger bone. The finger bone is long and curved, which is suggestive of an arboreal lifestyle, but I am not convinced that it belonged to *Orrorin*. It may be better assigned to the monkey from the site. The humerus, however, is definitely *Orrorin*'s. It has very strongly developed crests for elbow flexor muscles, which are typically powerful in climbers. So *Orrorin* was probably also adept in the trees and on the ground. In addition, *Orrorin* shows the same evidence of canine reduction that we see in *Sahelanthropus*, although we cannot be sure if the canine in *Orrorin* is male. Overall, I think the evidence is good that *Orrorin* is also an early hominin.

THE *ARDIPITHECUS* QUESTION

Ardipithecus ramidus is a well-known hominin from 4.4-million-year-old deposits in Ethiopia, described by Tim White from the University of California, Berkeley and his colleagues. Most researchers view *Ardipithecus ramidus* as a hominin, with evidence of canine reduction, a forward-placed and downward-facing foramen magnum and bones below the neck generally suggestive of a primitive form of bipedalism with retained arboreal capabilities. When I was a student I learned that bipeds can only be bipeds if they have an adducted big toe, that is, a big toe pressed up against the other toes, unlike the abducted, grasping big toes of all other primates, which function like a thumb. In bipeds, big toes function as a sort of springboard that allows our lower leg muscles to propel the body forward efficiently. It is the last part of the body to leave the ground in humans.

Our big toes are greatly enlarged compared with those of great apes, and we have a large muscle attached to it, the flexor hallucis longus, which goes only to the big toe, whereas this same muscle is shared among several toes in great apes.

Well, *Ardipithecus ramidus* proves us wrong. *Ardipithecus ramidus* has an abducted big toe, as in great apes and all primates other than australopithecines and humans. This finding completely changes our ideas about how bipedalism evolved. It is apparently possible to be a biped and to retain a grasping big toe, which is of obvious usefulness in climbing. Some researchers think that because of the grasping big toe and other primitive features, *Ardipithecus ramidus* is not a hominin at all but rather an African ape ancestor. I do not subscribe to this view. In fact, the problem with the idea of grasping big toes not being allowed in hominins is what I tell my students is the process-before-pattern problem. When you put process before pattern you impose a set of restrictions on the way you can interpret the evidence. If you are certain that a biped cannot have an abducted big toe because of some preconception about function and evolution, you will miss the incredible mosaic that is *Ardipithecus*. In evolutionary biology, as in a crime scene investigation, pattern must come first. Meticulous collection of data comes before interpretation. The data have to provide the support for a hypothesis, not the other way around. If you have a preconception, for example, that molars or brains can only get bigger over time, you will ignore or dismiss data that may reflect the real and different story. So for all of you budding scientists, let the data take you where it will and do not impose restrictions on them. I will return to this point in a bit.

When *Ardipithecus ramidus* was first named, it was actually referred to as *Australopithecus ramidus*. The authors decided to create a new genus, *Ardipithecus*, after realizing that it lacked an important characteristic found in both *Australopithecus* and *Homo* (this was before the skeleton was discovered). All hominins after *Ardipithecus* have expanded, or molarized, milk molars, but in *Ardipithecus* they look like chimp milk molars. I was actually in on the naming of *Ardipithecus* to some degree, having urged Tim White to create a new genus before someone else did. I am sure I was not alone. That same logic holds for "*Ardipithecus kadabba*." The quotation marks indicate that I do not think it is really *Ardipithecus*. From this point on I just call it "Kadabba."

Kadabba was described by my colleague Yohannes Hailie-Selassie. I got to know Yohannes when he was a graduate student at Berkeley, and I know that he is an excellent researcher, but I have to disagree with him on this point. Kadabba lacks an important feature found in *Ardipithecus* and all later hominines: a nonsectorial canine/premolar complex. In Kadabba it is sectorial.

A sectorial canine/premolar complex is the norm among anthropoids and results from male anthropoids having enlarged canines. Typically, the back of the upper canine is honed, or sharpened, by rubbing against the front of the lower first premolar. This feature is most strongly developed in male monkeys and apes, but it occurs to some extent in females as well. Upper canines in anthropoids with well-developed honing can be as sharp as a razor and are used in male encounters rather than being a tool for processing food. Male baboons, which have the largest canines relative to body mass, have been known to kill each other in fights. However, the theory is that large canines are mostly for show—what primatologists call threat displays—to prevent rather than promote conflict. With smaller and less impressive canines, the sectorial canine/premolar complex in Kadabba is not like that of baboons; it is more like what we see in Miocene and living great apes, especially chimps and bonobos. To my mind, this means that Kadabba is not in the same genus as *Ardipithecus ramidus* (figure 9.3).

Whether Kadabba is actually another species of *Orrorin* or *Sahelanthropus* or is a new genus, it is quite different and more primitive than *Ardipithecus ramidus*. Kadabba is complicated for a number of reasons. First, it is an amalgam of specimens from several localities dated to between about 5.2 and 5.8 million years ago. The most important specimens are a toe bone and the canine/premolar complex that I just described. The canine and premolar are 5.6 to 5.8 million years old, whereas the toe bone is 5.2 million years old, but there is no way to be sure that the teeth and the toe bone are from the same animal. The toe bone is long and curved. The joint that comprises the ball of the foot, which anatomists call the metatarsophalangeal joint, is tilted upward on this toe bone, which is what

FIGURE 9.3. Canines of chimpanzees, Kadabba, and *Ardipithecus*. From left to right, the canines of a male chimpanzee, a female chimpanzee, Kadabba, and *Ardipithecus ramidus*. Kadabba's canines are similar to those of the female chimp and lack the further reduction that unites *A. ramidus* with *Australopithecus*. If the Kadabba specimen is male, then the canines show some reduction compared with those of male chimps (as does *Ouranopithecus*), but much less than later hominins. (Middle two images courtesy of Yohannes Haileselassie. Right image © and courtesy of Gen Suwa and Tim White, from Suwa et al. 2009. Reprinted with permission from AAAS. Left image by author.)

we see in bipeds. When you stand on tiptoe, you are bending your foot between the long bones of the foot, the metatarsals, and the toes. Because we bend at the foot between these joints, the joints are tilted in order to allow for this bending.

One thing that is certain, however, is that the canine and premolar exclude Kadabba from *Ardipithecus*, at least from a strict taxonomic perspective. Who knows about the toe bone? In a word, *Ardipithecus* is a taxonomic problem. To repeat, because it is important, if two species are placed in a single genus but one shares a derived character with another genus, we call that genus a paraphyletic taxon. It is the same problem we had when we used to lump all great apes into the pongids and humans into hominids. African apes share characters with *Homo* and *Australopithecus* that *Pongo* lacks, so *Pongo* cannot be in the same taxon as the African apes.

Stated another way, African apes and humans have to be united to the exclusion of *Pongo*. *Ardipithecus ramidus* shares a canine/premolar complex without honing with *Australopithecus* and *Homo* that Kadabba lacks. It is exactly the same reason that *A. ramidus* cannot be *Australopithecus*, as originally proposed: because it lacks expanded deciduous (milk) molars.

Ardipithecus ramidus is a well-defined taxon and it is most probably a hominin with an unexpected mosaic of chimp-like and humanlike features. It has a chimp-sized brain and an elongated, chimp-like face, but reduced male canines, a forward-placed foramen magnum, and features of the lower limbs that indicate this animal was bipedal. Its abducted big toe strongly suggests it was still skilled getting around in the trees.

Kadabba was identified as *Ardipithecus* because the describers saw it as a chronospecies of *Ardipithecus*. A chronospecies is a model that puts fossils in a time line of evolving species based on the progressive development of morphology. Kadabba is considered a chronospecies and ancestor of *Ardipithecus ramidus* because it appears to be transitioning from large apelike canines to smaller hominin canines. Again, this is putting process before pattern. The pattern tells us that Kadabba lacks characters found in *Ardipithecus*, *Australopithecus*, and *Homo* (namely, a nonsectorial canine/premolar complex). It may be the ancestor of *Ardipithecus*, but this is no more likely than *Orrorin* or *Sahelanthropus* as ancestors of *Ardipithecus*.

The difference lies in the way I interpret the evidence of Kadabba and *A. ramidus* and the way researchers who described those species interpret it. Ultimately this disparity represents a philosophical or epistemological (theory of knowledge) difference between people like me, who put evidence, or pattern, before process, and those who use a processual that is, a process-first, approach. I do not argue that it is unreasonable to expect that there is a predictable process that would explain the progression from larger to smaller canines in early hominins as they transition from being great-ape-like to modern humanlike. It makes sense, but we do not know what that process actually was. We have no evidence of a specific process, only the pattern. In my view an expectation of a certain process, like canine reduction, does not justify placing fossils in a row representing chronospecies. Based on the decrease in brain size and facial/dental size, we used to describe a straight line going from *Australopithecus* to modern humans, but we now know that the pattern and process are infinitely more complicated than this simple scenario. The tree of life is a bush with many twigs, many experiments in form.

The same researchers who use a process to make sense of a pattern, as opposed to the reverse, are those who commonly cite the frequency of homoplasy to explain morphology that does not "fit." Homoplasy, or parallel evolution, does occur frequently, but it cannot be used as a stand-by excuse to dismiss evidence. In the case of *Ardipithecus*, the authors concluded that it must have evolved from a quadruped of some sort, but not from a suspensory ape similar to living or Miocene apes. The evidence for this conclusion is weak at best. Humans share a very large number of features with great apes, as described at the beginning of this book, which only make sense if we evolved from a common ancestor that spent most of its time in the trees, sitting on or hanging from branches, with a backbone positioned vertically. Our mobile shoulders and wrists, our hinge-like elbows, our barrel chests, our shorter lumbar spines, and our orthograde posture are all shared with great apes. If this multitude of attributes did not evolve as a result of our shared ancestry with an arboreal, orthograde, suspensory ape, where did it come from? From a quadruped monkey-like taxon such as *Ekembo*? If *Ardipithecus* evolved from an *Ekembo*-like quadruped, then dozens if not hundreds of attributes shared among the great apes and humans must have evolved in parallel at least four times (in humans, chimps, gorillas, and orangs). This strikes me as unlikely in the extreme. In the analysis of *Ardipithecus*, the late Miocene fossil record of apes is virtually ignored. Not being able to read minds, I cannot fathom why this would be except as another example of process over pattern. Why would one ignore a vast amount of evidence of modern great-ape morphology in Eurasia and suggest an unbelievable amount of homoplasy in the hominoid skeleton except in the worldview that everything must have happened in Africa, as Darwin predicted (although with the caveat expressed by Darwin that *Dryopithecus* from Europe may be an African ape and human ancestor)?

Hominines first appeared in Europe 12.5 million years ago, proliferated into a number of genera and species, and persisted in Europe until about 9 million years ago. During this time nothing of a convincing nature attests to the presence of hominines in Africa. Then, 2 million years after they vanish from the fossil record of Europe, hominines appear in Africa in the form of hominins (human ancestors). I would love to be able to fill that 2-million-year-long gap.

THE "MISSING LINK"

So, from what kind of ape did humans evolve? Or did we even evolve from an ape as opposed to a proto-ape like *Ekembo*? We used to think that one of the middle or late Miocene apes with thickly enameled teeth, such as *Kenyapithecus* or *Ramapithecus*, was the ancestor of hominins, and in fact at one point they were even included together with *Homo* in hominids to the exclusion of living apes. We know now the story is very different and much more complex, given the many discoveries of the last twenty years in Asia, Europe, and Africa.

In their discussion of the spectacular discoveries of *Ardipithecus ramidus*, Tim White, Owen Lovejoy of Case Western University, and colleagues emphasize the degree to which it was unlike chimpanzees or other apes. There is no doubt that *Ardipithecus* and, by extension, humans did not evolve from chimpanzees. We both evolved from a common ancestor that was neither a chimp nor a human. However, this does not exclude the possibility that one of us (either chimps or humans) could more closely resemble our common ancestor. Chimps share many features with australopiths, such as chimp-sized brains; elongated faces with long, chimp-like premaxilla; long, low braincases; and molars with more flared sides. *Ardipithecus* has chimp-like, but also gorilla-like, abducted big toes and limb proportions intermediate between those of humans and African apes. Many other features of the limbs—such as the shoulders, elbows, and hand and foot bones—are also features found in both African apes and australopiths.

Of course, chimps and modern humans also share more DNA in common with each other than either do with gorillas. But we also share bone attributes with chimps and gorillas. As I noted earlier in this book, humans and chimpanzees are the only primates in which all four upper front teeth (incisors) are wide. In other anthropoids, only the middle two incisors are wide. Humans (including fossil humans) and African apes share a number of limb attributes not found in other primates, and these are clues about how bipedalism may have evolved.

Rather than assuming that the common ancestor of chimps and humans could not look similar to a chimp because both have evolved over the last 7 million years or more, let's look at the evidence. Put a chimp, an australopith, and a human in a lineup and the witness will pick the human as the odd one out every time. It is very difficult to deny that australopiths, loosely defined to include *Ardipithecus*, more closely resembled chimps than humans. So this raises the question, why humans have evolved from our common ancestor so much more than chimps? Or, is it that humans have evolved at a "normal" rate whereas chimps have changed very slowly? The simplest hypothesis is that both have evolved at the same rate, but the evidence indicates otherwise and deserves an explanation. Do we really know that all primates evolve at the same rate? Of course not. There is much more about the process of evolution that we do not know than there is that we do.

There is a phenomenon here that demands further investigation. Humans clearly have diverged to a greater degree from our common ancestor than have chimps. Our brains, faces, teeth, and anatomy below the neck have evolved more than those of apes, chimps in particular. I am not saying that chimps are living fossils, not having evolved at all since their divergence from humans, but I do not see any alternative to the idea that their morphological evolution has been much less extensive than that of humans since the divergence.

Perhaps this discrepancy in degree of evolution is not something that is special about either chimps or humans but is instead a side effect of the process of evolution. It may have to do with the numbers of humans and chimps that have lived since the divergence. Humans are fairly ubiquitous in the fossil record, whereas there is virtually no evidence of chimps in the fossil record until about 545,000 years ago, and even then we have just a small handful of teeth and nothing else until the present. This discrepancy does suggest that population sizes of ancient chimps were much smaller than those of ancient humans and that they had less diverse ranges and more restricted niches. By at least 3.5 million years ago, our ancestors were spread out from Ethiopia to South Africa, and by 1.8 million years ago we had already dispersed to Asia and Europe from Africa. In comparison with most other catarrhines (Old World monkeys and apes) in the fossil record, hominins are plentiful and diverse. We were in evolutionary terms a big success, as we are today, being far more numerous than all other primates alive combined.

It is a bit of an oversimplification to say that, as some have suggested, we have not found chimp or gorilla fossils simply because they lived in forests, which have acidic soils that prevent bone preservation. All early humans until at least 3 million years ago inhabited forested ecological settings most similar to those inhabited by many chimpanzees today, yet human bones survived. Nearly all the Miocene apes described in this book also inhabited forests. It is more likely that the ancestors of African apes were just plain rare. Smaller population numbers means less genetic diversity, fewer mutations, and less new genetic material for natural selection to work with. If African apes remained in the forests of equatorial Africa in relatively small numbers, it is quite natural to suppose that their morphology would not have changed much, given their small numbers and relatively stable ecological circumstances.

On the other hand, small population sizes favor the random or chance processes of evolution, genetic drift, and gene flow. In drift, new genes are introduced or eliminated from populations by random events, such as a small subset of a population moving to a new location, which we call the founder effect. If the population is small enough, the removal of certain individuals from the parent group and their founding of a new group might well lead to genetic differentiation, especially if they are prevented from interbreeding by some natural barrier, such as a river. Gene flow occurs when populations disperse and interbreed, introducing new genetic material into the group. Both of these effects are felt more strongly in smaller populations, in which the effects of the loss or gain of genetic material is stronger than in large populations, in which the movement of genes goes more or less unnoticed. This may explain in part why there is relatively little morphological diversity within chimpanzees as a group but quite a bit of genetic diversity. The recent discovery of large amounts of genetic diversity in all nonhuman hominoids has in fact led to the naming of new species of chimpanzee and gorilla and even new genera of gibbons, although not all specialists agree with these new names (see chapter 3).

Hominins, on the other hand, are found first in more forested conditions, but not wet tropical forests or mountain forests, like most gorillas, bonobos, or West African chimps. They are found in drier forests with mixed ecological signals (usually forests, woodlands with trees spaced farther apart, and some indications of open country), which are similar, in fact, to the forests in Eurasia in which we find apes, as well as to those in which we find some chimps, such as in Gombe National Park, Tanzania (where Jane Goodall studied her chimps). I think that some hominines dispersing into Africa from Europe were already adapted to a combination of forest and open-country conditions, conditions they had encountered in Europe. The first hominines to disperse into Africa probably favored habitats to which they had adapted in Europe: mildly seasonal forests. Two of the lineages of this ancestral African hominine—chimps and gorillas—remained in the relatively stable and predictable forests. As the habitat dried out and became more mosaic, some hominines that had evolved in Europe under more seasonal conditions had the capacity to expand their ranges into the more mosaic environments. These became the ancestors of the hominins.

It would be nice if we had fossil gorillas and chimps the same age as *Sahelanthropus*, *Orrorin*, and Kadabba, but we do not, so we are forced to speculate based on a combination of fossils and living animals. One difficulty in teasing out the real events that led to the origin of humans is the rapid evolution of humans. One of my first projects after my dissertation was to assess the relations of great apes and humans using *Australopithecus* and Miocene apes instead of modern humans as the taxa to compare with living great apes. *Australopithecus* retains many more features of our common ancestor with African apes than we do, so I was trying to eliminate the "noise" from 3 million years of evolution subsequent to the appearance of *Australopithecus afarensis* (figure 9.4). The result was some of the first evidence (1992) based on fossil morphology to support the molecular conclusions that chimps and humans are sister taxa. I want to take that logic back further, into the Miocene

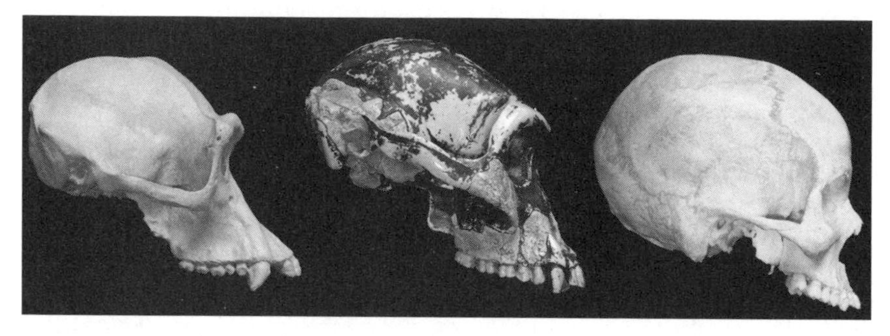

FIGURE 9.4. Skulls of a chimp, *Australopithecus afarensis*, and a human. Which one does not belong? Clearly, the human (*right*) has a larger brain case and smaller face than either the chimp (*left*) or australopithecine (*middle*). (Images by author.)

and look at the fossil evidence to help us understand relationships between African apes and humans and the origin of the human lineage.

When we carry out a phylogenetic analysis, as described earlier in this book, we can place taxa in a given sequence independent of geography or any assumption of a process other than evolution. The specific process that produced the pattern reveals itself to the intuition of the researcher by the pattern. The pattern of evidence from the Miocene indicates that there was a transformation over time from the more monkey-like *Ekembo* to the essentially modern ape-like fossil apes such as *Hispanopithecus* and *Rudapithecus*. *Afropithecus*, from the late early Miocene, suggests what the initial changes in feeding adaptation may have been like (toward broader diets, including difficult-to-extract foods). *Nacholapithecus*, from the middle Miocene, with its enlarged forelimb joints, suggests the earliest occurrence of more forelimb-dominated behavior. By the late Miocene, apes, the vast majority of which are found in Eurasia, were essentially modern, suspensory, and orthograde. They matured more slowly and had large brains relative to their body mass. What remained was for one of them to disperse into Africa between 9 and 10 million years ago, or perhaps earlier, and radiate into the ancestors of the living African apes and humans.

DOWN FROM THE TREES OR UP FROM THE GROUND?

If the late Miocene European hominines were suspensory animals, does this mean that bipedalism evolved from an ancestor that was suspensory? Not everyone thinks so. For example, the idea proposed by White and colleagues that human bipedalism evolved from some form of palmigrade quadrupedalism, such as that reconstructed for *Ekembo* and seen in most living non-hominoid primates. If the ancestor of humans had walked on all fours, with its palms facing downward like a baboon's, there would be no explanation for the observation that humans retain the dozens of attributes from the neck down that link us with all apes. Since Thomas Henry Huxley in the nineteenth century, anthropologists have recognized the shared features of the thorax, shoulders, arms, legs, hands, and feet in African apes and humans. All of them are plausibly related to upright posture and swinging beneath the branches of trees.

But enough about the palmigrade quadrupedalism hypothesis. Nearly all researchers since Huxley's time agree with him that humans evolved from some sort of a suspensory animal. There are two main scenarios associated with this hypothesis. One is that humans essentially went directly from being orthograde in the trees to being orthograde on the ground. Researchers point to the orangutan. Orangs are probably the most orthograde of the great apes, spending most of their time in the trees holding onto tree trunks and hanging from large branches. They will also walk bipedally on large branches, sort of. Orangs will walk on large, horizontal branches while stabilizing themselves by grabbing branches above their heads. This manner of walking is known as forelimb-assisted bipedalism, and it has been proposed as a precursor to terrestrial bipedalism. My colleagues Brian Richmond, David Strait, and I have called it hindlimb-assisted suspension. The point is that this behavior is biomechanically completely unlike human bipedalism on the ground. Orangs are much less adept on the ground than African apes. Their arms are much longer and not at all suited to support their body mass other than in a hanging position. Their feet are also less suited to hold up the body, being more like hands than feet in many ways, which is why they were long called quadrumanous, or four-handed. Most of the time when they move on the ground, orangs adopt a posture known as fist-walking. Their hands are so long and their fingers so curved that they have to curl their hands into fists so that the backs of the hands touch the ground. This is very inefficient and they don't cover much distance in this way. There are some interesting muscular similarities between orangs and humans around the hips, which were first noted by a modern pioneer in comparative hominoid anatomy and positional behavior, Jack Stern of Stony Brook. However, these are few in comparison with the number of characteristics shared among African apes and humans.

The orthograde hypothesis tells some of the story. It explains the characters shared among all hominoids related to orthogrady and suspension. But it does not explain the transformation from a body adapted to a highly arboreal lifestyle to one adapted to an exclusively terrestrial one. The second part of the story helps to explain the transition to terrestrial locomotion.

I mentioned earlier that African apes and humans share attributes of the shoulders, elbows, wrists, hands, and feet not found in orangutans. The shoulders are rotated outward and the elbows are more stable and have a strong hinge. The wrists have eight bones instead of nine, with the centrale being fused to the scaphoid; these remain two separate bones in nearly all other primates. There is also less movement between most of the bones of the wrists and hands in hominines than in orangs. The hands are shorter and the fingers are less curved. The feet are more stable and function as better platforms for walking than in orangs. All of these traits are plausibly related biomechanically to knuckle-walking.

A knuckle-walker holds the hand in alignment with the forearm and bends the fingers at the second joint of the digits, so that the middle phalanges are flat on the ground (figure 9.5). There are differences between chimps and gorillas, most of which are related to size, such as whether all four phalanges or just a few touch the ground, but the mechanics are almost identical. To prevent the wrist and palm from buckling under the weight of the front of the body, many of the joints involved have special stabilizing features. These

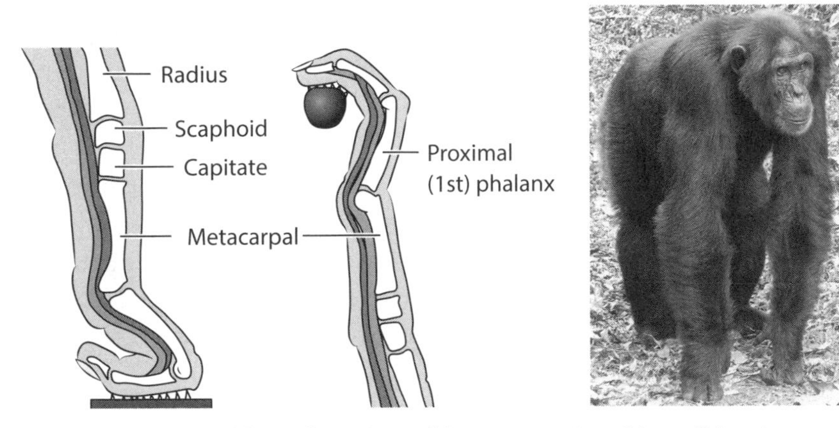

FIGURE 9.5. Knuckle-walking in a chimpanzee. Knuckle-walking is an excellent strategy for maintaining competence both on the ground and in the trees. The hands retain long metacarpals and curved phalanges for arboreal activities but also have specializations, including locking mechanisms to stabilize the wrist, palm, and fingers in knuckle-walking. The left drawing shows a chimp's hand in the knuckle-walking posture, and the right drawing shows it in the suspensory posture. (Left image modified from Filler, A. G, 2007. *The Upright Ape: A New Origin of the Species*. Wayne, NJ, New Page Books. Right image courtesy of John Mitani.)

include bone ridges that lock the joints in place as well as projections of bone that muscles use to leverage the wrist and hand into the proper position. All in all it works quite well. Knuckle-walkers are excellent terrestrial quadrupeds without being palmigrade and excellent climbers without being quadrumanous.

The funny thing is that although humans are not knuckle-walkers, we share with chimps and gorillas nearly all the attributes that distinguish them from orangs. The stabilizing attributes of the wrist and hand that make knuckle-walking possible have been repurposed to make our hands stable for manipulating objects and perhaps for carrying them also. The stable feet of knuckle-walkers are an even more obvious preadaptation for the even more stable feet of hominins. There is no evidence that any known hominin was a knuckle-walker, and some evidence, such as the position of the foramen magnum in *Sahelanthropus* and the femur of *Orrorin*, to indicate that they were not. However, the retention of many attributes of the limbs between knuckle-walkers and humans and the simple fact that both are forms of terrestrial locomotion make the knuckle-walking hypothesis compelling in my view. It is the second phase in the development of bipedalism, after a more generalized orthograde arboreal phase. It is also the most *parsimonious* hypothesis.

What do I mean by "parsimonious"? We know that chimps and humans are most closely related to each other and that gorillas branched off earlier. Since chimps and gorillas are knuckle-walkers, there are two possibilities. Either knuckle-walking evolved once, in the common ancestor of chimps and gorillas (which is of course also the common ancestor with humans), or knuckle-walking evolved independently in gorillas and chimps. Put another way, if the common ancestor of African apes and humans was a knuckle-walker, then gorillas, which branched off first, retained this adaptation as did the common ancestor of chimps and humans. No changes so far. Then, following the chimp-human divergence, chimps retained knuckle-walking and humans became bipeds. One change.

The alternative, if humans do not have a knuckle-walking ancestry, is that the common ancestor of African apes and humans was not a knuckle-walker but rather, say, an arboreal orthograde climber. Gorillas branched off first and evolved knuckle walking while the chimp-human ancestor retained orthograde climbing. One change occurs, knuckle-walking in gorillas. Then, chimps and humans split, and knuckle-walking evolved again, independently, in chimpanzees, while bipedalism evolved in humans, that is, two more changes, for a total of three changes as opposed to one. Of course, by "changes" I mean transformations from one form of positional behavior to another that required many anatomical changes, probably over some period of time. Nevertheless the knuckle-walking hypothesis is more parsimonious; in other words, it is simpler and consistent with the evidence. The orthograde climber hypothesis is more complicated and less consistent with the evidence of the anatomical similarities among African apes and humans.

One thing the record of Miocene apes does not reveal clearly is the reason for the origin of bipedalism, one of the first things to occur in human evolution. There is much debate about this development because there is little direct evidence.

WHY BIPEDALISM EVOLVED

The tool-making hypothesis. Early researchers including Darwin speculated that bipedalism evolved to free the hands from the task of locomotion and transfer them to toolmaking. This is logical but almost surely either over-simplified or just plain incorrect. The first evidence of bipedalism in hominins is at least as old as 7 million years ago, if we accept the evidence of *Sahelanthropus*. Certainly by 4 million years ago bipedalism was fully developed in taxa like *Australopithecus*. Yet the oldest evidence of the manufacture and use of durable tools made from stone instead of bones, antlers, or plant parts is only about 3.2 million years old. And we have to wait another 1.5 million years or so before we start to see well-made, symmetrical stone tools like hand axes. In addition, the hand in early hominins like *Australopithecus* has a comparatively small and weak thumb, as in the great apes, unlike the longer and more powerful thumbs of later humans. There is little evidence in the hands to suggest that australopithecines, or for that matter, the earliest members of the genus *Homo*, were much better at making stone tools than are chimpanzees. Of course, it is likely that, just as chimps do, the earliest hominin bipeds made tools from perishable materials that are not preserved in the fossil record. Perhaps our toolmaking skills benefitted from some of the changes that were happening in our hands, such as the reduction of the length of the fingers. But we have no direct evidence that the earliest bipedal hominins were making tools that enhanced their chances of survival. Certainly the fact that our ancestors began to walk on two legs rather than four made it possible for toolmaking to evolve, but it was not caused by selection for skillful stone-tool making.

The carrying hypothesis. Another hypothesis is that bipedalism evolved to facilitate carrying things, such as food or babies. If you are moving around between clumps of forest and more open country in search of food and safe resting places, it would be a valuable survival asset to be able to carry food and infants that cannot walk on their own. Many other primates, including all the great apes, move around with babies, which usually ride on the adults' backs or clutch their chests and bellies but also are carried in their arms. So what was so important about carrying in protohominins that lead to selection for bipedalism? Perhaps a means of safely covering larger distances, extending home ranges, and thus increasing the opportunities to find enough food to survive and feed your young? Again, this is all speculative.

The thermoregulation hypothesis. Some researchers have suggested that bipedalism may be related to spending more time in open country, where there is less protection from the sun. The surface of the body directly facing the sun, the top of the head and shoulders, is smaller than in a quadruped, whose entire back faces the sun. So it may be a matter of thermoregulation, maintaining a constant body temperature. Humans are in fact very good at thermoregulation. Less densely distributed hair helps by removing some insulation and provides a smooth surface for sweat to accumulate and then evaporate, an efficient mechanism for removing heat from the body. We also have sweat glands that are more efficient at cooling our bodies than any other primate's. Humans are also excellent endurance movers, both walking and running. We can't outrun many animals over short distances, but if we keep on their tails over kilometers, eventually they slow down. This was a traditional method of hunting among hunter-gathers in southern Africa, such as the !Kung people of the Kalihari (also formerly known as Bushmen, which is generally considered to be a pejorative term today). However, it is likely that this form of endurance running, or even walking, evolved later in human evolution, as our bodies became more streamlined (narrow at the hips with long limbs), increasing both surface area for evaporation and stride length for more efficient bipedalism. This did not happen until about 1.8 million years ago, much later than the origin of bipedalism.

The roots-and-tubers hypothesis. Bipedalism may have evolved as a response to selection pressure to make the gathering of foods, in particular fruits and roots and tubers (underground storage organs) more efficient. Bushes or small trees not easily climbable may have been more accessible to a biped standing on two feet with outstretched arms. Using two hands to unearth and carry roots and other underground storage organs may also have conferred an advantage to early hominins.

Of course, there is one more possibility: chance. It is hard to imagine, but theoretically possible, that the first biped arose as a random event. In my view that possibility is vanishingly small given the number of changes to the skeleton, from the head to the toes, needed to be an effective biped.

It could be that aspects of all these hypotheses are correct. They are not necessarily mutually exclusive. There was an element of chance, in that mutations arising randomly conferred some adaptive advantage to evolving bipeds. But selection accounts for the development of these traits. Perhaps removing the onus of locomotion from the upper limbs was useful in allowing protohominins to make and use tools made from perishable materials that do not preserve well in the fossil record. That, coupled with carrying, makes sense, especially if you want to carry around your tools. Carrying food and babies to keep them safe from predators may have been a factor, but there probably had to be other reasons, since most primates carry their babies around (for that matter, even crocodiles carry their babies—in their mouths). Thermoregulation and endurance may have been factors but probably would only have become strong in association with the origin of modern human body form, such as we find in *Homo ergaster* at around 1.8 million years ago.

LAST APE STANDING

The ancestors of gorillas developed dental adaptations to allow them to exploit highly fibrous foods in times when their preferred source of calories and nutrients, ripe fruit, was less available. To this day, gorillas eat sweet, ripe fruits when they can get them but fall back on THV (terrestrial herbaceous vegetation) when they need to. This strategy has allowed them to survive shortages of their preferred foods, and it would continue to work for them were it not for humans encroaching on their territories and even poaching them for food and souvenirs.

Gorillas are endangered (mountain gorillas) or critically endangered (lowland gorillas). This sounds like a local issue for the people who live in proximity to gorillas, but it is in fact a global issue. People are motivated by a strong desire to survive and to feed and provide for their children. Sadly, too many find gorillas to be easy targets for sources of protein (so called bush meat) and sellable parts (souvenirs, alleged aphrodisiacs, and orphaned babies) and their habitats to be converted to farm land. The citizens of developed nations are quick to condemn Africans for the decline of African ape populations, without realizing that we actually contribute directly to this phenomenon by failing to fully reciprocate for the resources we extract from that continent and by providing markets for contraband. The continued survival of gorillas will depend on cooperation from the African and developed nations.

Back in the late Miocene, another population was probably more adaptable, able to exploit a wide range of resources in a more varied environment. This was the population that gave rise to the last common ancestor of chimpanzees and humans, commonly known as the Missing Link. What chimps eat and where they live are not that different from the preferences of gorillas in many instances, but they entirely lack the special adaptations that gorillas have for extracting energy and nutrients from THV. Instead, chimps are more flexible in their ability to exploit resources in a greater diversity of environments, and they seem to have adapted to shortages of food by developing strategies to procure more energy from animals. Chimps are the only apes that routinely hunt for termites and ants and use specially designed tools to do so. They fashion spears to hunt small animals such as bush babies. Male chimps cooperate in hunting smaller individuals (females and young) of monkeys like the red colobus, and infant antelopes and pigs. This may well protect them from the uncertainties of the environment and may be one aspect of their fallback strategy. But unlike gorillas that rely on THV, chimps rely on food that is more challenging to acquire, and doing so may represent selection for greater cognitive capacity. It is tempting to see the selection for the intellectual capacity to find embedded foods as a precursor to the development of the human mind, but researchers are divided as to whether we can say that there is any real difference in cognitive capacity among the great apes.

Chimps, it is sad to say, are suffering the same fate as gorillas. Both species (bonobos, *Pan paniscus*, and common chimpanzees, *Pan troglodytes*) are also on the endangered species list. The future for great apes is grim. Deforestation, whether for harvesting products like exotic woods or simply clearing land for farming, is eating up their habitats at alarming rates. The bush meat and animal trophy trades are also taking a terrible toll. We are all implicated in this massacre. In many cases local people are desperate to feed their families, and we in the West provide the markets for forest products, whether legal or not. Industrialized countries invest in development programs that encourage deforestation or have other negative effects on the environment. Unless there is a concerted effort internationally to eliminate these markets and there are major changes in governance to provide local people in the countries in which great apes live with more resources to support their families, I do not hold out much hope for the great apes, or most primates for that matter.

One thing the great apes, and primates in general, have that may work in their favor is our emotional attachment to them. People are outraged by stories of their demise in the wild and especially of their mistreatment in our care. The use of primates in medical research is vilified by many, and where it does occur, the labs are under heavy security and a veil of secrecy. I have even heard rumors of chimps on various university campuses that no one talks about for fear of reprisals. Even before conservationism entered our collective consciousness, we have long had a fascination with apes. They look more like us than other animals, and the more we learn about them, the more we see how much they also act like us. The point of this book is to explain what we share with our primate cousins and our great ape brothers and sisters and how those similarities came to be. We are killing our great ape brethren, but maybe, just maybe, we might find a way to keep them around for generations to come. Otherwise, we will be the last ape standing.

POSTSCRIPT

As I was finishing this book I had an amazing experience that took me back to the earliest days of my career. I have always wanted to work in France on the fossil record of ape evolution. The first fossil ape recognized as such, *Dryopithecus*, was discovered in France and published in 1856. Darwin knew of and commented on this discovery, as I have mentioned. Very little has been discovered in France relevant to ape evolution since the late part of the nineteenth century, but the quality of the fossils found in the "little Pyrenees" of France, that is, the foothills, is good. I always hoped to be able to find more. In 2014 I found out that the original site of the discovery of *Dryopithecus* had been rediscovered by colleagues from the Natural History Museum in Toulouse, which is about 80 kilometers north of the site.

In the summer of 2014 I went to visit the property near St. Gaudens, in the department of the Haute Garonne, France, with my colleague Guillaume Fleury from Toulouse. It had been rediscovered by another colleague, Yves Laurent, who could not come with us on that day. Guillaume and I arrived at the place and pressed the buzzer, which we hoped would alert someone at the home about a hundred meters away. It is a big property. We had tried to call ahead but to no avail. Still, we wanted to try our luck. No one showed up, so we went to the neighbors to see if someone was home. The neighbor was not sure, but when we turned around to try again, there she was, the proprietor of what is, we were hoping, one of the most historic sites in paleoanthropology.

The person who greeted us was an elegant woman of a certain age, as we say in France. While kindly and gracious, there was an odd sadness to her that we soon understood. The way she put it was, "Yesterday I buried my husband," the French idiom for "Yesterday was my husband's funeral." Her husband had died earlier in the week after a long illness, and we could not have arrived at a worse time. We apologized for the intrusion and turned to leave but she insisted that we stay, saying "It would change nothing." Reluctantly we walked up the long driveway to her house and she ushered us inside. She showed us the file that our colleague had left with her husband, who was enthusiastic about renewed excavations on his property. We, of course, were familiar with all the documents she was showing us, but obviously, given the circumstances, we sat patiently as she recounted what we already knew about our colleague's previous visit. After a while she invited us to have a look at the back of the property, where there is a high quarry wall, the remnants of the clay quarry that had been in the family for multiple generations.

Her husband's great grandfather was the owner of the quarry at the time that the first jaw of *Dryopithecus* was discovered. Talk about your six degrees of separation! He found the mandible and recognized that it was something interesting, though he knew not what. As far as we can tell, he brought it to an acquaintance of some erudition, either a physician or a pharmacist (in the day and in the countryside the two could have been the same), a Monsieur Fontan. The rest of the story has been documented many times. Monsieur Fontan showed the specimen to Edouard Lartet, who recognized it immediately as an ape and published it under the name *Dryopithecus fontani*, making specific comparisons with African apes.

We had met the widow of the man whose great grandfather had held in his hand for the first time ever a fossil hominine. It was a pleasure and an honor to spend that couple of hours with this wonderful lady. Her son and grandchildren showed up during our time there, and the son even pitched in to help us clear a path to a part of the quarry wall. The day after his father's funeral! I think it was a tribute to his father and his interest in the history of the locale.

We may be able to go back to the site in the future and pick up where Lartet left off, but this time with more modern methods. If we are lucky we will be able to date the site using state-of-the-art methods such as paleomagnetism, and we will hopefully be able to collect more fossils. Very few fossils are known from the site, so the more we find, the more we will be able to learn about the ecology of this early hominine. The geochemistry and geomorphology of this site, specialized approaches in geology, will reveal to us details of the ecology and landscape in which these earliest of hominines lived. A return to this historic site could not be a better place for me to be at this point in my career.

The next day Guillaume, Yves, and I went to Sansan, a major paleontological site that is the reference for the middle fauna of Europe and the place at which the first fossil primate ever described was found. My pal Lartet was again responsible. As I mentioned, he was able to see beyond the prejudices of the day and the place and saw *Pliopithecus* for what it is, a fossil catarrhine. We scrambled up to the quarry and also found Monsieur Lartet's small farmhouse. To most people it would be seen as it is, a ruin, but to me it was a place of pilgrimage. Plus ça change, plus c'est la même chose. Merci Monsieur Lartet.

ACKNOWLEDGMENTS

Early in my career I met some amazing people to whom I am immensely grateful for help and guidance. Eric Delson, a major figure in paleoanthropology, has been tremendously supportive since I met him in 1980 as an undergraduate collecting data at the American Museum of Natural History in New York. Being my first time behind the scenes at a museum, I will always remember Eric Delson's generosity in helping me find my way, which he has continued to do to this day.

In 1986 I needed to get away from teaching to finish my thesis, and I applied for a fellowship at the Smithsonian Institution. For some reason, Rick Potts, the recently hired curator of paleoanthropology, supported my application and I got the job off the short list, after someone declined. Thank you whoever you are. At the Smithsonian I got to know many phenomenal researchers in paleoanthropology and related fields and endeavors, including John Gurche, Jenny Clark, Kay Behrensmeyer, Hans Sues, Dick Thorington, Kathleen Gordon, Mike Petraglia, Tom Plummer, and Liz Bailey, not to mention the numerous researchers who came to the museum in the three years that I was there. Rick and many others I got to know at the Smithsonian remain close friends to this day.

I was at the right place at the right time when in 1988 I published a paper on the finger bones from Rudabánya, which was noticed by Alan Walker. Alan and colleague Mark Teaford had 200-plus finger and toe bones to analyze from their spectacular site in Kenya, the Kaswanga Primate Site at Rusinga. While I was still at the Smithsonian, Alan hired me to teach anatomy at Johns Hopkins and work on those bones and other things, such as the endocast of the *Homo ergaster* adolescent skeleton from Nariokotome (Kenya). That kept me going for another year until I was hired at Toronto. In addition to the chapter on the brain of the Nariokotome boy that I wrote with Alan, I described the 200-plus finger and toe bones from Kaswanga, which Alan later described as the longest and most boring article ever written, as I noted earlier.

Alan was the recipient of a Macarthur Award (commonly known as the genius award), which he used very generously to support the early careers of many researchers, including mine. At Hopkins I was aglow in the light of an amazing group of researchers that along with Alan included Mark Teaford, Joan Richtsmeier, Ken Rose, Chris Ruff, Pat Shipman, Nikos Solunias, and Dave Weishample. The students there, when I was a postdoc, are among today's leaders in the field: Carol Ward, Chris Beard, and Larry Witmer, to name a few. Being in that atmosphere, both at the Smithsonian and at Hopkins, changed my life. It solidified in my mind that what I was doing was interesting and worth doing.

I am extremely grateful to Alison Kalett at Princeton University Press for asking me to write this book. It takes as a starting point an article I wrote for *Scientific American* under the guidance of Kate Wong, to whom I am also grateful. Many thanks as well to Ann Downer-Hazell, who helped with her editing skills to get my points across and who asked some excellent questions, the answers to which I hope enhance this book.

My great friends, the "Magnificent Seven" (you know who you are), have made me feel at home and very happy in this field for many years. Finally, I can't thank enough Dana Bovee and our children André, Alyx, and Mara for being so amazing and tolerant of my frequent absences and for even more frequently agreeing to allow me to drag them along. Dana has been my partner for most of our lives. In addition to all of that, we have been collaborators at Rudabánya for ten seasons, where Dana, an archeologist, designed the documentation protocol that allows us to reconstruct the site down to the nearest centimeter. I dedicate this book to her.

NOTES

INTRODUCTION

1. We know today from genetic and anatomical evidence that tarsiers are more closely related to anthropoids than they are to prosimians, but because they behave like many prosimians (being nocturnal, insectivorous, and solitary), they are classified informally as prosimians. More formally, primates are split into strepsirhines and haplorhines. Tarsiers are included with anthropoids as haplorhines.

CHAPTER 1

1. In the interest of full disclosure I should say that not everyone agrees with this view. Terry Harrison, the colleague I mentioned earlier, does not accept the evidence for loss of a tail in *Ekembo*. Coccygeal vertebrae are small and fragile, and in fact we do not have one that we can definitively identify for *Ekembo*. But we do have the bottom of the sacrum, to which the coccyx attaches, and it looks like one that would have had a coccyx attached to it. We do have a coccyx of a later-occurring ape, *Nacholapithecus*, which is the first completely unambiguous evidence of the loss of the tail in apes.

CHAPTER 2

1. A type specimen is the object on which the species is based. All species are supposed to have a type specimen that researchers can refer to when evaluating new discoveries. The name, or nomen, is essentially attached to the type, and wherever the type goes, so goes the nomen. If the type is placed in a previously named species after further research reveals that it in fact should not be a separate species, the species name becomes a synonym of the previously named taxon. This is how we manage to some degree at least to maintain order in the tree of life. Interestingly, the oldest recognized species (those named by Linnaeus) do not have types, because this convention was not in place at the time. Among the original Linnaean taxa for which no type was designated is *Homo sapiens*. Some have suggested that the great Swedish naturalist himself should be the type. Any other volunteers?

CHAPTER 3

1. There are actually two Děvínská Nová Ves sites. One is a fissure fill site (like a sinkhole) that contains a remarkable collection of skeletons, including the primitive catarrhine *Epipliopithecus*. The *Griphopithecus* site is a few hundred meters away and completely distinct geologically. It is a massive sand deposit, testifying to the presence of a once large and impressive river.

CHAPTER 4

1. *Equatorius* was named in 1999 based on a reanalysis of the fossils from Maboko Island following the discovery of a partial skeleton from Kipsarimon. Some authors continue to use the taxon *Kenyapithecus* for the Maboko specimens.

2. This mandible is universally attributed to *Proconsul* (now reclassified as *Ekembo*) today.

3. As noted earlier, today "hominid" refers to the great apes and humans, in recognition of the extremely close genetic similarities among them. At the time, "hominid" was reserved for taxa more closely related to humans than to any other taxon.

CHAPTER 5

1. Charles Darwin. 1871. *The Descent of Man*. London: Murray, p. 135.

CHAPTER 6

1. At the risk of making my many friends who work in natural history museums unhappy, I do have to say that I have always been uncomfortable that alongside displays of dinosaurs and the world's ocean life and rain forests, there are halls devoted to Native American, African, South American, Australian, Oceanean, Asian, and other indigenous cultures. Natural history has a long record of lumping ecology and paleontology with anthropology, which resulted in the construction of these different displays.

2. Funnily enough, I tried to be Pilbeam's student, twice. I was not admitted to either program. No hard feelings. It eventually worked out for me.

3. Tracy has long held a different point of view about the evolution of knuckle-walking as me. Not counting *Sivapithecus,* I think that knuckle-walking probably evolved once, in the common ancestor of the African apes and humans. Tracy thinks that it evolved independently in chimps and gorillas. This is a very important point that I will get back to in chapter 9 of this book, because it has a bearing on how we reconstruct the behavior of the last common ancestor of African apes and humans. I commend Tracy for holding her ground, even though her supervisor (me) and an influential researcher who was a member of her dissertation committee disagreed with her (and we still do). It is a good lesson for students: defend your position, if it is defensible, and don't blindly accept everything your elders tell you; doing so will give you credibility and respect.

4. Traditional Chinese medicine is receiving more and more attention from Western medical practitioners. It is widely regarded as effective, but there is little "science" behind it. Currently there is a large effort in China to combine the methodology of western pharmacology with traditional medicine to establish the actual effectiveness in treating ailments. It is an interesting merging of traditions, but also an effort that if successful would open up a large market for Chinese traditional remedies in the West. Perhaps one day researchers will find the active ingredients in seahorses, rhino horn, musk deer musk pods, and the skins, bones, teeth, and claws of tigers, all endangered and all common ingredients in the huge Chinese traditional medicine industry. In so doing, there is hope that it may be possible to avoid the extermination of all of these pharmaceutical target animals. Among the alternate sources of these ingredients (which also include domestication and farming), perhaps someone will find out what is good for you in the fossils of *Gigantopithecus*, thus saving a few for researchers to ponder.

5. "Don't look a gift horse in the mouth" refers to the fact that the height of a horse's teeth is an indication of its age. A horse with worn teeth is old, but because it is a gift, it is impolite to complain.

6. The islands have not moved since the retreat of the ice sheets. Instead, the ocean has returned. During the glacial maxima of the Pleistocene so much water was locked up in glaciers, which were miles thick in places, that sea levels around the globe were significantly lower than today. One could walk from France to England, and the Southeast Asian mainland was much larger, incorporating most of the islands that make up Malaysia and Indonesia today.

CHAPTER 7

1. Gábor died in 2012 at the age of 81. He found most of the primates at Rudabánya from 1967 until I began excavations there in 1997, and even then, with more than a dozen students on site, he still managed to find the best primate fossils. *Anapithecus hernyaki* was named in honor of Gábor in 1976.

2. My use of the taxon *Dryopithecus* differs from that of some authors. I include in this taxon two other genera, *Pierolapithecus* and *Anoiapithecus*, from the Vallés Penédés in the Catalonia province of Barcelona, Spain. While there may be more than one genus actually present among these taxa, they are all, in my view, closely related to each other and broadly ancestral to the later occurring dryopithecins *Rudapithecus* and *Hispanopithecus*, which I discuss in more detail in chapter 8.

3. The radius attaches to the scaphoid and lunate bones on the thumb side, and the ulna attaches to the triquetrum and pisiform on the pinky side.

REFERENCES

INTRODUCTION

Darwin, C. 1859. *On the Origin of Species by Means of Natural Selection or the Preservation of Favored Races in the Struggle for Life*. London, Murray.
Fleagle, J. G. 1999. *Primate Adaptation and Evolution*. New York, Academic Press.
Huxley, T. 1863. *Evidence as to Man's Place in Nature*. London, Williams and Norgate.
Martin, R. D. 1990. *Primate Origins and Evolution*. Princeton, NJ, Princeton University Press.

CHAPTERS 1–2

Begun, D. R. 2003. "Planet of the apes." *Scientific American* 289: 74–83.
———. (2013. "The Miocene hominoid radiations." In: *A Companion to Paleoanthropology*, edited by D. R. Begun, 397–416. New York, Blackwell Publishing.
Begun, D. R., and L. Kordos. 1993. "Revision of *Dryopithecus brancoi* SCHLOSSER, 1901 based on the fossil hominoid material from Rudabánya." *Journal of Human Evolution* 25: 271–285.
Guatelli-Steinberg, D. 2010. "'Growing planes': Incremental growth layers in the dental enamel of human ancestors." In: *A Companion to Biological Anthropology*, edited by C. S. Larsen, 485–500. Chichester, UK, Wiley-Blackwell.
Harrison, T. 2013. "Catarrhine origins." In: *A Companion to Paleoanthropology*, edited by D. R. Begun, 376–396. Hoboken, NJ, Wiley-Blackwell.
MacLatchy, L. 2004. "The oldest ape." *Evolutionary Anthropology: Issues, News, and Reviews* 13(3): 90–103.
Walker, A., and P. Shipman. 2005. *The Ape in the Tree: An Intellectual and Natural History of* Proconsul. Cambridge, MA, Belknap Press of Harvard University Press.
Walker, A., and M. F. Teaford. 1989. "The hunt for *Proconsul*." *Scientific American* 260: 76–82.

CHAPTERS 3–4

Begun, D. R., et al. 2003. "The Çandır hominoid locality: Implications for the timing and pattern of hominoid dispersal events." *Courier Forschungsinstitut Senckenberg* 240: 251–265.
Leakey, M., and A. Walker. 1997. "*Afropithecus*: Function and phylogeny." In: *Function, Phylogeny and Fossils: Miocene Hominoid Evolution and Adaptations*, edited by D. R. Begun, C. V. Ward, and M. D. Rose, 225–239. New York, Plenum.
Nakatsukasa, M., and Y. Kunimatsu. 2009. "*Nacholapithecus* and its importance for understanding hominoid evolution." *Evolutionary Anthropology: Issues, News, and Reviews* 18(3): 103–119.
Ward, S. C., and D. L. Duren. 2002. "Middle and late Miocene African hominoids." In: *The Primate Fossil Record*, edited by W. Hartwig, 385–397. Cambridge, Cambridge University Press.

CHAPTERS 5–6

Ciochon, R. L., et al. 1990. *Other Origins: The Search for the Giant Ape in Human Prehistory* New York, Bantam Books.

Kelley, J. 2002. "The hominoid radiation in Asia." In: *The Primate Fossil Record*, edited by W. Hartwig, 369–384. Cambridge, Cambridge University Press.

Kelley, J., and F. Gao. 2012. "Juvenile hominoid cranium from the late Miocene if southern China and hominoid diversity in Asia." *Proceedings of the National Academy of Sciences* 109(18): 6882–6885.

Nargolwalla, M. C. 2009. "Eurasian middle and late Miocene hominoid paleobioogeography and the geographic oroigins of the Hominindae." PhD diss. University of Toronto.

CHAPTER 7

Begun, D. R., et al. 2012. "European Miocene hominids and the origin of the African ape and human clade." *Evolutionary Anthropology: Issues, News, and Reviews* 21(1): 10–23.

Kordos, L., and D. R. Begun. 2002. "Rudabánya: A Late Miocene subtropical swamp deposit with evidence of the origin of the African apes and humans." *Evolutionary Anthropology: Issues, News, and Reviews* 11(2): 45–57.

CHAPTER 8

Alba, D. M. 2012. "Fossil apes from the Vallès-Penedès basin." *Evolutionary Anthropology: Issues, News, and Reviews* 21(6): 254–269.

Bonis, L. de, and G. D. Koufos. 1994. "Our ancestors' ancestor: *Ouranopithecus* is a Greek link in human ancestry." *Evolutionary Anthropology: Issues, News, and Reviews*: 75–83.

Delson, E. 1986. "An anthropoid enigma: Historical introduction to the study of *Oreopithecus bambolii*." *Journal of Human Evolution* 15(7): 523–531.

Jerison, H. J. 1973. *The Evolution of Brain and Intelligence*. New York, Academic Press.

CHAPTER 9

Brunet, M., et al. 2002. "A new hominid from the Upper Miocene of Chad, Central Africa." *Nature* 418: 145–151.

Haile-Selassie, Y. 2001. "Late Miocene hominids from the Middle Awash, Ethiopia." *Nature* 412: 178–181.

Senut, B., et al. 2001. "First hominid from the Miocene (Lukeino formation, Kenya)." *Comptes Rendus de l'Académie des sciences, Sciences de la terre et des planètes* 332: 137–144.

Simpson, S. W. 2013. "Before Australopithecus." In: *A Companion to Paleoanthropology*, edited by D. R. Begun, 417–433. Hoboken, NJ, Wiley-Blackwell.

Suwa, G., et al. 2009. "The *Ardipithecus ramidus* skull and its implications for hominid origins." *Science* 326(5949): 68, 68e1–68e7.

White, T. D., et al. 2009. "*Ardipithecus ramidus* and the paleobiology of early hominids." *Science* 326(5949): 64, 75–86.

INDEX